# SEEDS
# OF
# RESISTANCE

## THE FIGHT TO SAVE
## OUR FOOD SUPPLY

# BY MARK SCHAPIRO

HOT BOOKS

Hot Books may be purchased in bulk at special discounts for sales promotion, corporate gifts, fund-raising, or educational purposes. Special editions can also be created to specifi-cations. For details, contact the Special Sales Department, Skyhorse Publishing, 307 West 36th Street, 11th Floor, New York, NY 10018 or info@skyhorsepublishing.com.

Hot Books® and Skyhorse Publishing® are registered trademarks of Skyhorse Publishing, Inc., a Delaware corporation.

Visit our website at www.hotbookspress.com.

10 9 8 7 6 5 4 3 2 1

Library of Congress Cataloging-in-Publication Data is available on file.

Cover design by Brian Peterson

Print ISBN: 978-1-5107-0576-0
Ebook ISBN: 978-1-5107-0580-7

Printed in the United States of America

# CONTENTS

# FOREWORD

## BY DAVID TALBOT

The world is burning, and yet the firelight illuminates the way out. The times are dire, even catastrophic. Nonetheless we can sense a grand awakening, a growing realization all around the globe that "people have the power, to dream, to rule, to wrestle the world from fools" in the prophetic words of Patti Smith.

But in order to rouse ourselves from the nightmares that hold us in their grip, we need to know more about the forces that bedevil us— the structures of power that profit from humanity's exploitation and from that of the earth. That's the impetus behind Hot Books, a series that seeks to expose the dark operations of power and to light the way forward.

Skyhorse publisher Tony Lyons and I started Hot Books in 2015 because we believe that books can make a difference. Since then the Hot Books series has shined a light on the cruel reign of racism and police violence in Baltimore (D. Watkins' *The Beast Side*); the poisoning of U.S. soldiers by their own environmentally reckless commanding officers (Joseph Hickman's *The Burn Pits*); the urgent need to hold U.S. officials accountable for their criminal actions during the war on terror (Rachel Gordon's *American Nuremberg*); the covert manipulation of the media by intelligence agencies (Nicholas Schou's *Spooked*);

the rise of a rape culture on campus (Kirby Dick and Amy Ziering's *The Hunting Ground*); the insidious demonizing of Muslims in the media and Washington (Arsalan Iftikhar's *Scapegoats*); the crackdown on whistleblowers who know the government's dirty secrets (Mark Hertsgaard's *Bravehearts*); the disastrous policies of the liberal elite that led to the triumph of Trump (Chris Hedges' *Unspeakable*); the American wastelands that gave rise to this dark reign (Alexander Zaitchik's *The Gilded Rage*); the energy titans and their political servants who are threatening human survival (Dick Russell's *Horsemen of the Apocalypse*); the utilization of authoritarian tactics by Donald Trump that threaten to erode American democracy (Brian Klaas's *The Despot's Apprentice*); the capture, torture, and detention of the first "high-value target" captured by the CIA after 9/11 (Joseph Hickman and John Kiriakou's *The Convenient Terrorist*); the deportation of American veterans (J Malcolm Garcia's *Without a Country*); and the ways in which our elections have failed, and continue to fail, their billing as model democracy (Steven Rosenfeld's *Democracy Betrayed*).

And the series continues, going where few publishers dare. You hold in your hands the latest offering in our series—and in some way, the most elemental. Nothing is more essential to human existence on our planet than seeds, the tiny pods that sustain us all. And yet the basis of our food supply is coming under increasing stress—by climate change and the corporate and political forces that are pushing us to the brink of survival. The Trump presidency is only aggravating this crisis in our food chain, with its refusal to believe in science or to heed urgent environmental alarms. Fortunately, as Mark Schapiro explores in *Seeds of Resistance*, a global movement of farmers and grassroots guardians of our food supply has sprung up to fight the corporate greed and political ignorance that endangers us all. While much of Schapiro's book is a wake-up call about the growing threats to the origins of our existence, his book is also deeply inspiring, documenting how unsung heroes around the planet continue to nurture and protect the agricultural cradles of life.

Hot Books are more condensed than standard-length books. They're packed with provocative information and points of view that mainstream publishers usually shy from. Hot Books are meant not just to stir readers' thinking, but to stir trouble.

Hot Books authors follow the blazing path of such legendary muckrakers and troublemakers as Upton Sinclair, Lincoln Steffens, Rachel Carson, Jane Jacobs, Jessica Mitford, I.F. Stone and Seymour Hersh. The magazines and newspapers that once provided a forum for this deep and dangerous journalism have shrunk in number and available resources. Hot Books aims to fill this crucial gap.

American journalism has become increasingly digitized and commodified. If the news isn't fake, it's usually shallow. But there's a growing hunger for information that is both credible and undiluted by corporate filters.

A publishing series with this intensity cannot keep burning in a vacuum. Hot Books needs a culture of equally passionate readers. Please spread the word about these titles—encourage your bookstores to carry them, post comments about them in online stores and forums, persuade your book clubs, schools, political groups and community organizations to read them and invite the authors to speak.

It's time to go beyond packaged news and propaganda. It's time for Hot Books . . . journalism without borders.

# INTRODUCTION

# SEEDS & CANARIES

**SEEDS ARE EVERYWHERE.** Blowing in grains of pollen, secreted in hard shells underneath the surface of the earth, cached inside of fruits and nuts. They come in all forms: round, oblong, square, rectangular, blue, brown, yellow, green, and all the colors in between. Each one is a packet of kinetic energy, a genetic call and response, a story yet to be written. Add three basic ingredients—soil, water, and sun—and the story unfolds. First the seed breaks through its shell, then it pushes its roots into the soil in search of water, and finally it sends its stem upward toward the sun.

This is the primordial process that supports life on the planet. Emerging on branches, tucked into flowers, or sometimes multitasking as roots, every seed holds the promise of becoming food. And each contains information about the conditions in which they evolved, the adaptations that enabled them to survive, and the human beings who cultivated and nurtured them.

Around ten thousand years ago, humans figured out how to stop wandering and plant seeds in the ground, which meant they no longer had to go on endless treks to hunt for food. Thus began agriculture, and the deeply mutual interplay between humans and seeds. This

1

included keeping an eye out for the seeds that could survive shifting weather patterns and deliver the best looking, tastiest, and most nutritious food. Centuries before Charles Darwin came along to give it a name, farmers played with the forces of evolution. Year after year, season after season, they would select the strongest seeds to nurture and pass along to their descendants, continuing the story.

Now those ancient seed narratives are being interrupted, intercepted and driven by climate change into unexpected new terrain, presenting a challenge to how we grow food.

**THREE DECADES AGO,** I was one of a small group of journalists writing about the impact of narrowing genetic diversity in seeds.[1] But none of us could have anticipated the alarming forces that have put this process on warp speed. Just as the never-before-seen growing conditions call out for a broader selection of seeds, there's been a steady shrinking of options as control over them slips to barely a handful of global companies, all of which specialize in substances that kill plants and other organisms.

At the same time, a global movement of farmers, scientists, activists, and people concerned about healthy food are fighting for a more resilient agriculture and autonomy over their seeds.

We'll begin with the battle over seeds and the emergence of seeds at the center of the battle over the future of our food. In the second part, I'll introduce you to some of the "seed warriors"—the farmers, scientists, activists, librarians, explorers, indigenous leaders, and botanical wizards who are on the frontlines of the struggle for control of the earth's most fundamental resource: our seeds.

**BEFORE WE BEGIN,** allow me to introduce myself, your narrator. As I mentioned, I've been reporting on and writing about the environment since the early 1980s, when I began my career at an investigative nonprofit called the Center for Investigative Reporting. I was interested in revealing the impact of abuses of power on our health and on our natural environment. Since then, I've chased illegal loggers down Amazonian back-roads, trudged through oil splattered rocks on an Atlantic beach in the aftermath of an oil spill, spied on trucks loaded with illegal pesticides headed to ports for shipments overseas, plodded through drought-parched farm fields, waded into rivers that crossed frontiers squabbling over water rights, and had my blood tested to find out how many toxic chemicals are running through my—and likely your—veins.

Over the years, I've been asked many times, in one form or another, "How do you keep doing this kind of work and not throw yourself off a cliff?" Metaphorically speaking, of course. I never quite know how to answer. I was certainly inspired by Rachel Carson and other journalists early in the history of environmental journalism to reveal who is responsible for degrading our common natural resources and threatening the health of humans and other organisms for private benefit. To document unfair and unjust abuses of power is a first step toward change.

As to where that impulse comes from, it helped that my father was a neuro-endocrinologist, which meant he studied the cascade of hormones that contribute to a brain's awakening from infancy to adulthood. His findings on infant rats' responses to various stimuli—sound, light, taste—would later contribute to how we understand the developing human brain.[2] In my father's laboratory I "helped" out as only a twelve-year-old could—by holding a stopwatch and timing frantic rats trying to make their way out of a maze, to measure their intelligence. Aside from somewhat comical moments watching determined and often lost rats, I learned that if you follow the trail of evidence, you can get to the story behind the action, whether it's the endocrine system of a rat or the depredations that imperil the ecological networks which sustain our natural world.

In this sense, journalists' methods are not that different from scientists'—following strands of evidence to probe into the established set of "facts," and examining assumptions. Paradigms hold until new evidence comes along to shake them.[3]

There was, no doubt, another seed planted during that formative time, and it has to do with the powers of fate. On a flight from the Peruvian city of Cusco to Lima, the plane in which both my parents were flying failed to clear the Andes and crashed. They were on their way home after a two-week trip built around my father's presentation at a scientific conference. From the peaks of the Andes came an "act of God," a terrible twist of fate. I was fourteen; my two brothers, Erik and Seth, were eleven and eight. Also killed in addition to our parents, Shawn and Lorraine Schapiro, were a delegation of Quechua community leaders headed to the capital and a group of American high school students headed home to Connecticut after a summer abroad program.

There was little to "understand" about such a life-changing event, as tragic as it was for us and all the other American and Peruvian families impacted. There was no one to blame.

It was only years later I realized something that should be obvious but is not often considered in the same breath: trauma wrought by fate and trauma wrought by acts of intention are profoundly different. The former are rolls of the dice; there's no way to avoid them. Accidents do happen. In the environmental realm, most of the erosion of eco-systems, and the individuals and communities they support, are not acts of God, but acts of humans. With enough information pried loose, these players and their machinations can be revealed. And if they can be revealed, they can be confronted and, perhaps, stopped.

**WHICH BRINGS US** back to seeds. Like all environmental stories, start with a seed and you quickly end up in the realms of money and power—who has it, and who's struggling to gain or regain it. And, of course, you end up in the realm of science.

The latest science suggests that plants, including those of our major food crops, are engaged in a continuous interplay of responses with the environment in which they're planted.[4] That environment is changing; climatic disruptions are accelerating. The number of seed companies is declining, and the spectrum of seeds shrinking. The group of people involved in fighting for their seeds, and a more just and healthy food system, is expanding. Old assumptions of how we grow food are falling. New paradigms are emerging. It's a time of profound vitality and volatility in the seed realm, with high stakes for all of us who care about our health, the planet's health, and the food we eat.

As powerful forces circle round the ground-zero ingredient of our food, one thing is becoming clear: a seed is never just a seed. Seeds are the canaries on our climate disrupted planet.

They're emitting strong signals. Let's read them.

# CHAPTER 1

# A SEED CHRONICLE FORETOLD

**A SEED STORY,** like life, starts small and gets bigger.

In the mid-1990s, a letter arrived at a simple adobe-style office on a dusty lot on the outskirts of Tucson, Arizona. The site, headquarters of Native Seeds/SEARCH, an organization which saves seeds native to the southwest, is little more than a couple of garden plots and a refrigerator and freezer filled with indigenous seeds. Here you can find seeds that have inherited characteristics dating back as far as three-thousand years.

The letter came from a lawyer for the food company Frito Lay, known for its processed snack foods. It warned that one of Native Seeds/SEARCH's best-selling products, Indian Parched Corn, was in violation of the Frito Lay trademark. Parched corn are chunky salted corn kernels that have been prepared for centuries by Hopi, Apache, and other tribes in Arizona and New Mexico. (A reminder to the non-botanical: kernels are in fact seeds, so when you eat corn you're eating a corn seed.) The company lawyer demanded that Native Seeds/SEARCH immediately "cease and desist" using language equating the native corn with two words that had been trademarked by the company—Corn Nuts™.

The command arrived like a tragi-comic bolt from the big-food universe. A multi-billion-dollar company was objecting to language used by a tiny seed saving enterprise promoting a seed that has been around for hundreds, if not thousands, of years, long before Frito Lay was a phantasm in the mind of the first mass marketer of the potato chip, Herman Lay, in 1932. The parched corn comes in four and eight ounce bags with yellow, blue, and red kernels, reflecting the diversity of some eight hundred different corn varieties, cultivated from the American southwest down to southern Mexico. It's not as if the inhabitants of this region had to go to Frito Lay to figure out how to make the tasty morsels made from dried corn kernels. They've been doing it for centuries, through an age-old practice of boiling ("parching") the kernels, drying them and adding salt.

The main supplier to Native Seeds/SEARCH is the Santa Ana Pueblo in neighboring New Mexico, located about twenty-five miles from Albuquerque, home to a federally recognized tribe which dates its presence in the area to at least the sixteenth century. Talavai Denipah-Cook, a twenty-three-year-old Hopi woman who works as an ecologist on tribal lands and grew up in the Ohkay Owingeh Pueblo, just north of Santa Fe, recalled to me in 2017 that she used to enjoy eating parched corn regularly as a kid. "They're crunchy and delicious," she said. They taste, well, like nuts made from corn.

No matter how well disguised the decorative packaging or how thoroughly they've been put through the additive grinder, the genes of the corn used by Frito Lay emanate from seeds that are relatives of the seeds that have been cultivated by indigenous Americans (North and South) for thousands of years. Generation upon generation of the region's farmers ensured that the seeds' genetic pool was preserved, and thus provided the foundational raw material for the company's processed food concoctions.

The seed group could have responded by requesting that Frito Lay, itself a subsidiary of gigantic PepsiCo, express its gratitude for the centuries of corn cultivation by the peoples of what is now Mexico and the American Southwest.

That didn't happen.

Without the resources to fight one of the world's biggest food companies, Native Seeds/SEARCH changed the disputed wording. "Parched Corn" packets now state, simply and directly, "Parched without oil for a healthy, crunchy, and uniquely Southwestern snack."

"They seemed to come from a different continent, a different world," said Laura Jones, the group's acting director, as she recalled the incident in 2016, while ushering me into the group's refrigerated seed vault one hot summer morning in Tucson. She opened a foot-thick steel door, and we walked into the refreshingly cold air. There, on steel shelves, I gazed upon the genetic history of the American southwest. Piled into plastic canisters were seeds of every description—small ragged-edged oblong seeds, square-ish, round and trapezoidal seeds, future beans, squash, peas, corn, a desert wheat, and many others in dark browns, light browns, blues, blacks, yellows and whites. "This is our palette of genetic diversity, our insurance that we'll be able to survive and thrive here in the southwest," Jones said. As the world comes to resemble Arizona—hot and dry—those seeds are becoming ever more invaluable.

**THOUGH A LEGAL** conflict was averted, the larger question at the heart of this dispute remains: who controls our seeds? We members of the public have a major stake in the answers to that question, which will determine the quality, nutrition, and healthfulness of the food that we eat, especially as our planet's ecosystems are being degraded and unsettled.

The financial wizard played by Brad Pitt in the film *The Big Short* advises his young acolytes that "seeds are . . . the new currency." Indeed, while few were looking, seeds became one of the hottest properties in the corporate market. American and European seed companies have been buccaneering their way through the world's seeds, picking and choosing the ones to patent and put into mass production. This means they have retained exclusive rights to resources that have been in the public domain for millennia.

By 2018, after a frenzy of mergers and acquisitions, just three companies controlled more than half of all seed revenues, and a growing percentage of the living germplasm embedded in those seeds. The primary business for all three, now fused into globe-stretching merged companies—DowDuPont, Bayer-Monsanto, and Syngenta-ChemChina—is not seeds, but agricultural chemicals. The combination of chemical and seed companies is giving rise to seeds that are born addicted to chemicals for their survival—entire generations full of crack-baby seeds.

One major result has been the accelerating disappearance of seed diversity at just the time when we require a broader genetic spectrum to adapt to the volatile impacts of our changing climate. Changes in growing conditions are convulsing the planet's food-growing lands. The changes are coming more quickly, and less predictably, than our ability to breed seeds can respond. The time by which adaptability to new conditions must occur is shortening.[1] We can no longer wait the average five-to-ten- or even fifteen-year time span needed to breed a new variety.

Not only is it getting hotter and drier, tumult in the atmosphere is delivering conditions that we can no longer predict, which means that agriculture has to respond both to the current set of conditions, and to new and more volatile conditions, which may be just a season or two ahead. Climate change is heightening the stakes in our search for seeds that can adapt to an accelerating pace of unknowns. "There's no precedent at all for what we're going to see," George Frisvold, a professor of Agricultural and Resource Economics at the University of Arizona, told me, when I visited him in his office on the other side of town from Native Seeds/SEARCH in Tucson. Frisvold served on President Clinton's Council of Economic Advisers where, in the 1990s, he began trying to gauge the impact of climate change on our food system. Now, it seems, the concerns they had at the time are coming true like a chronicle foretold. "We don't really know how bad it's going to get."

A diversity of seeds is imperative to enable us to withstand this period of accelerating environmental stress. Threats loom. The conflict over Corn Nuts in many ways reflects the larger conflict over control of the seeds and thus our ability to respond to climate chaos as the world's most important public resource falls increasingly into private hands.

SECURITY OF OUR food supply is the big story, and its filled with many smaller stories along the way. Here's another one.

May 19, 2017 was the day that the inconceivable happened. In the far north of Norway during a record-breaking warm spring day, preceded by a record-breaking warm fall, rain fell instead of snow. The permafrost melted and sent water fifty feet into the entryway of a three-hundred-foot long mountain tunnel leading to the world's largest repository of seeds: the Svalbard Global Seed Vault. Close to a million seed samples are stored inside, representing the germplasm of more than four thousand distinct food crops, a seed cornucopia—including beans from the American Southwest, corn from Mexico, rice from Southeast Asia, potatoes from Peru, wheat from the Middle East, cowpeas from the horn of Africa, and pomegranates from Central Asia. The vault had been deemed impregnable, the safest imaginable place for hedging against the disappearance of genetic resources that are the source-pool for our present and future food. Suddenly the world's "failsafe" backup seemed exceedingly vulnerable.

The earth's most important genetic resources, protected by an entire mountain, were threatened by the direct impact of the earth's most significant man-made threat: climate change. It would be hard to find a more apt and ominous foreshadowing for the challenges ahead for the world's seeds.

The flood at Svalbard was a brutal reminder of the difference between a diversity of seeds in a frozen vault—whether in Norway or Tucson or any of the dozens of such facilities across the world—and a diversity of seeds out growing in the fields, responding and adapting to conditions that are changing more dramatically and decisively than ever before. The seed vault at Svalbard offers some reassurance that the resource key to human survival has an important, if imperfect, backstop to disappearing forever. But its inadequacy was also revealed. The tundra melted and with it the false sense of security we might have enjoyed in thinking about a vault inside a mountain—or a vault anywhere for that matter. There have been numerous studies documenting the ideal temperature at which seeds can be kept alive.[2] Keeping seeds alive in a jar or a Petri dish, though, is far different from keeping seeds alive when they are growing in a field, exposed to and interacting with the elements.

The Svalbard seeds were ultimately saved when an emergency Norwegian team was able to block the water in the entrance corridor before it reached them deeper inside the vault. The world breathed a sigh of relief. But outside the vault, in the fields of the earth, seeds face far greater threats than they do from a melting mountain, and these threats are not on the horizon. They are here, now. The converging forces of economic concentration and ecological disruption have far greater implications for the future of our food security than the ice-mountain meltdown at Svalbard.

**I VISITED A** seed vault for the first time in 1982, as a young journalist. There was a crisis in American farm country: farmers were being squeezed by a narrowing of dominant players in the global commodity markets and government policies favoring large farms. Family farmers were disappearing at nearly unprecedented rates, displaced from the land that many had farmed for generations. This was the first big shift in American agriculture. Family farms were being combined and consolidated into large commodity-oriented farms. (Within thirty years, by 2015, more than half of all Iowan farmland was owned by absentee landlords).

Genetic resources, it was said, were going with them. I'd never heard of genetic resources, but they sounded important. They certainly didn't seem like anything you could see, visit, or hold onto.

I finagled a magazine assignment—*'new phenomenon, seed banks!'*—and hopped an Amtrak train from Oakland, California across the sprawling flat Great Basin of Nevada and Utah, and into the Rocky Mountains. The National Seed Storage Laboratory in Fort Collins, Colorado was the first refrigerated seed storage facility in the United States. Located on the campus of Colorado State University and run by the US Department of Agriculture (USDA), it's where I learned what a genetic resource is: a seed.

Seeds contain genes accumulated over many seasons. From the most succulent fruit to the most bitter vegetable, every seed contains codes, braided into their DNA, that signal countless factors, including how much of each of those elements is needed to transform the seed into a plant, and if we're lucky, an edible, nutritious, and tasty one.

We see the ones that survive because of their particular combinations and ability to adapt. Those that can't survive are lost, often forever. Those that do survive have particular genetic combinations enabling them to change with the circumstances. At first domesticated agriculture was very much hit-or-miss, a mix of blind faith and pleas to the gods. But over time, the differing response of crops to water, sunshine, and soil started to make sense to the people who planted them. Patterns became clear. Every season became the test of a hypothesis. Maybe this or that seed will do better in the heat, or with diminished water, or fighting off a new threat. Harvests would prove or disprove the thesis.

Season after season this was the science—though it wasn't called that—which lays behind growing our own food. Which is why seeds are called "genetic resources"—they have the genes to survive and thrive, characteristics which, through intentional or accidental combinations, pass those characteristics onward. They are a resource borne from the fundamental Darwinian forces of ecological balance and brutal selection.

Humans, of course, have stored seeds for millennia, in barrels and urns, barns and basements, but the Fort Collins seed bank was the first outfit to create a "safe space" for seeds, intended not necessarily for planting but for preservation, to protect them from the threats outside. Two decades later, the director of the NSSL (since renamed the National Center for Genetic Resources Preservation), was among a group of distinguished scientists working with the Crop Diversity Trust to establish the Svalbard Seed Vault[3].

Inside a non-descript concrete block were thousands of seeds from every major food crop grown in North America. I recall the space as utilitarian—little more than row upon row of multi-colored and differently shaped seeds in jars and plastic pouches perched on steel shelves behind a walk-in freezer door—very much like those I would see decades later, on a smaller scale, in Tucson. The vibe was slightly medical and spooky sci-fi. I recall a sense of fascination with the place, but no sense of urgency; it seemed like a zoo for living seeds rather than a last-stand redoubt for a precious resource. For those running the institution, though, the intent was clear: to insure the primary ingredient of our food supply against future human and natural threats.

Looking back, it seems like an age of innocence. Back then, the primary threats to seed diversity were rapidly expanding cities, the slowly encroaching impacts of desertification, and the shift to cultivating commodity crops on ever-larger farms, which was leading to increased reliance on monocultures. Who could guess that within a decade dozens of formerly independent seed companies would be purchased by the world's largest chemical companies? Or that those companies would eliminate thousands of locally evolved seeds and favor the ones reliant on their chemicals? Or that genes from an unrelated organism would be inserted into germplasm to create genetically modified crops that would displace competing seeds? Or, critically, that all this would occur as the accumulation of greenhouse gases—barely a glimmer on most scientist's radar at the time—would threaten the status quo growing conditions on the earth's food-growing lands?

Farmers live with the consequences of these changing conditions every day. Each of the three essential elements that triggers a seed to become a plant—sun, water, and soil—are being altered profoundly by climate change. What we non-farmers might experience as another overly warm day, or perhaps even a welcome freak rainstorm, can for a farmer mean the difference between the life or death of their crops.

Despite the absurd and tragic denial of the facts by US President Donald Trump, one thing is clear: it's not getting any cooler. Rain is not falling when it used to, and when it does, it comes in increasingly intense and destructive bursts. The United Nation's International Panel on Climate Change, the leading international body of climate scientists, says that our current emissions scenario will lead to a three- to five-degree Fahrenheit global temperature increase by 2050. At current emission trajectories, US farm productivity will likely decline by 2.8 percent to 4.3 percent annually and drop to pre-1980 levels by 2050, according to a team of scientists centered at the University of Maryland in a study published by the National Academy of Sciences.[4]

One-sixth of all agricultural production is traded internationally, which means, as the USDA concluded, water and heat stresses are likely to have a severe impact on the "types and costs of food available for import."[5] We're already seeing the impacts: yields of tree crops are falling in California's Central Valley because it does not get cold enough to permit a metabolic slow-down in the winter before trees blossom with fruit in the spring.[6] Millions of acres are being fallowed each year in California, in the Southwestern states of Arizona, New Mexico, and Texas, and in the Midwestern wheat belt of Kansas and Nebraska, because there is not enough water to sustain crops. Fifty-six percent of the world's irrigated crops and 21 percent of rain-fed crops—a total of one-third of all food production—are already in areas subject to severe water stress, according to Ceres, a non-governmental organization (NGO) of financial professionals working with companies to address environmental risk.[7]

Heads up: here comes the desert to a neighborhood near you or a neighborhood where your food is grown. Every year another thirty million acres of food-growing lands are lost to desertification, says the United Nation's Food and Agriculture Organization.[8] Many of those in the direct path of desertification are in key food growing areas of the planet—including South Africa, Australia, Jordan, Peru, southwest China and, much closer to home for North Americans, southern California, much of Texas, large parts of Arizona and, south of the border, major farming regions of Mexico and Central America. As if that's not enough, ag journals like *Global Change Biology*, *Plant Disease*, and the *Journal of Insect Conservation* are full of reports of new plant and livestock pests and diseases following the rising heat northward. The biggest challenge for the United States and for the world is how to continue producing food in these turbulent conditions.

The USDA's alarm over the disruptive impacts of climate change on agriculture mounted during the latter years of the Obama administration—culminating with a report on the "profound threat" to food supply chains from climate-disruptions in the United States, and in the country's most important sources of food imports.[9]

"The geography of agriculture is changing," Margaret Walsh, chief author of the report, told me. About a year after I spoke with her, after Donald Trump assumed the presidency, the USDA issued a directive ordering its employees to remove the words "climate change" from the agency's vocabulary.[10] Alas, that semantic trick will do nothing to change the underlying reality.

Our food security hinges on the struggle now underway over who controls the earth's seeds as the ground shifts beneath our feet.

CHAPTER 2

# GENETIC VULNERABILITY: HOW WE GOT HERE

**IOWA, IN THE** summer of 1970, is where alarms were first triggered about the perils ahead for American agriculture. In July of that year, the green corn stalks on the McLain family farm were normal, about three feet tall, straight as arrows, vibrating in the hot breezes. In a couple of months, they'd be ready to harvest.

Twelve years later, I was standing with Fred McLain in his fields when he recalled that fateful summer, still vivid in his mind. He encouraged me to take in the site of the healthy corn farm surrounding us—the corn stalks fluttering in a light wind, marching seemingly forever over the family's 1,400 acre farm just off Highway 30 in the middle of the state. This, he said, was very much what his corn looked like back then, in July 1970, vigorous and healthy.

Then came August, when he began to see that something was amiss. He still recalled the sight like it was yesterday. "First, we began to notice little brown spots," he told me. "Then the ears turned brown and black." By the end of the month, he watched helplessly as one after another of the leaves were scarred with ink-like spots, growing bigger day by day. By October, when he gave up on the year's harvest and cleared his ravaged fields, black spores were sailing over the fields and towns of central Iowa like a plague of locusts.[1]

McLain was not alone during that distressing season: thousands of farmers across the United States experienced similar wipeouts of their crops, with experiences disturbingly like those described by McLain. The speed of the infestation was startling: over just one month, the blight spread from corn farms in the southeast, where it apparently originated in Florida or the Carolinas, across more than a thousand miles into the Midwest. In some parts of the country, as much as fifty percent of the corn crop was destroyed; nationwide, corn yields plunged fifteen percent, representing a huge loss to farmers of what was then, and still is, America's leading crop. Cattle eat corn, so prices for meat skyrocketed. Overall, the epidemic cost farmers billions of dollars in lost revenues, and seriously disrupted the supply chain for manufacturers who rely on corn as one of the main ingredients in processed foods.

The McLains' place, run by Fred, sixty-two at the time, and his son Frank, in his mid-thirties, was the first farm I'd ever visited. Frank had gone to college and decided to come back to his family's farm rather than head out to Chicago or Des Moines. I told him I lived in San Francisco. He shrugged, and said, wistfully, that he hadn't yet seen the ocean. I recall my incredulous response: "The ocean. Really?" He'd seen neither the Pacific nor the Atlantic. It had never occurred to me that someone roughly my own age would never have seen the defining feature of life on the coast, not to mention life on the planet. And then he asked me, in a friendly, jesting tone: "Have you ever been on a farm before?" I was in my late twenties. No, I hadn't ever been on a functioning farm before. That shut me up real fast. I was as profoundly distanced from the sources of my food as, I suspect, are most people. For him, the reality of those swaying fields of corn and soy and other crops he and his family grew were as fundamental a part of the landscape as the Pacific Ocean was to me. Each of us was missing something fundamental about life on earth. Frank would go on in the decades to come to see and enjoy the oceans. That was the beginning of my effort to try to understand the pressures on farmers that the McLains and many others face.

**SCIENTISTS SCRAMBLED TO** figure out how one disease could have traveled across America's corn fields so far so quickly. It didn't take long for them to identify the culprit. Major Goodman, who runs the Maize Breeding and Genetics Program at the University of North Carolina in Raleigh, explained to me years later that all the plants which died back then contained identical DNA emanating from the same parent seed. The corn blight, he said, was the result of American farmers using "a single nucleotide mutation in the genome . . . that was used by everybody." Translation: more than two-thirds of American corn fields were planted with the same hybrid strain of what was known as T male-sterile cytoplasm, or Texas cytoplasm,[2] the same corn seeds planted by the McLains. The hybrid corn, developed by two Texan plant breeders (hence the name), left male corn plants sterile. It eliminated the need for de-tasseling, the labor-intensive practice of removing corn tassels to exclude the prospect of cross-pollination, a practice that once offered many a Midwestern high school student a summer income.

The cytoplasm came along with another trait—susceptibility to the *Helminthorporium maydis*: a virulent corn fungus. When the fungus hit, it spread across America because all those corn stalks were, genetically speaking, the same plant. They were all highly vulnerable to the same fungus—a fate not unlike that which struck inbred royal families whose descendants grew progressively weaker as the gene pool of their parents shrank.

That devastating summer born of seed uniformity looms over the American food system like a Sword of Damocles.

**TO FIND A** way out of the spiraling disaster, panicked agronomists and plant geneticists flocked to a place far from Iowa—to the Sierra Norte mountains in the Mexican state of Oaxaca. Here in these mountains and valleys is corn's center of origin, its genetic home-base. Walk through the markets in the bustling Mexican city of Oaxaca, and you'll see corn in bounteous abundance—bins overflowing with corn kernels in every color: blue, red, purple, brown, and shades in between. Each have different characteristics, evolved and bred in ecological pockets within the mountain ranges and valleys to the north and south of Oaxaca. Mexico's National Commission for the Knowledge and Use of Biodiversity identifies sixty-five distinct maize landraces—the core parent lines that evolved over thousands of years within what is now Mexico—from which have sprung twenty-two thousand different corn varieties. (Of those, more than two thirds of US corn acreage is planted today with just four distinct varieties.) All that Mexican maize has a common ancestor, a wild corn plant called *teosinte*, the forefather of the corn we eat today. *Teosinte* isn't particularly palatable to humans—the cobs are small, tough, and stringy—but it does contain genetic treasures of response and resistance to a range of conditions and threats that have evolved over centuries. Among those characteristics were precisely what was needed—resistance to the corn leaf fungus.

The scientists moved rapidly to mix the Mexican corn seeds with existing US corn. By the next year, new seeds were in circulation and

Fred McLain's corn was on its way to recovery. He, like thousands of other farmers and millions of consumers, should be thankful to the Mexican farmers who had cultivated that corn for thousands of years, selecting their best seeds for planting with each new crop cycle. But, until recently, there was no official way to acknowledge the important contribution of such indigenously cultivated crop varieties to what we eat. It would take another three decades until that contribution was formally recognized—in the International Treaty on Plant Genetic Resources for Food and Agriculture. The United States finally ratified the treaty in 2016. While historic in providing global acknowledgement of the high value of genetic resources, by 2017, putting teeth into its provisions to compensate those responsible for keeping the genetic strains in thousands of seed varieties alive and thriving was in hot dispute.

In the treaty, 144 countries agreed to grant access to their publicly held germplasm in return for payments into a "co-benefits" fund that is supposed to be used to support biodiversity in resource origin countries, most of them located in a broad belt of land around the equator. The first has occurred—there are active exchanges underway from public gene banks, like the ones in Fort Collins and in the nine seed banks supported by the United Nations that are key to maintaining seeds in different ecological settings. Among those are the International Center for Tropical Agriculture, CIAT, with gene banks in Brazil and other tropical countries; CIMMYT, the International Maize and Wheat Improvement Center, headquartered in Mexico, a center of origin for many of our most familiar cereals and grains; and ICARDA, the International Center for Agricultural Research in Dry Areas, based, until recently, in Syria (and to which we'll return later). The Crop Trust in Svalbard is supposed to receive duplicates of the seeds in all these seed banks, to ensure there's a backup to the backups.

But the matter of who pays and how much continues to rage. As of 2017, little, if any, money had been paid into the fund. At the annual meeting in Rwanda that year of the treaty signatories the biggest controversy was over who should pay and how much, and how much access to public seed banks should be granted to private companies. A coalition of developing and EU countries, supported by European NGOs like the Swiss international watchdog group Public Eye, are advocating for a regular payment system that would ensure at least $50 million annually to promote seed diversity in center of origin countries, and compensate them for the invaluable task of sustaining such varieties over multiple generations for the ultimate benefit of all humankind. Thus far, however, efforts to establish anything stronger than a voluntary payment system have been strongly opposed by the major seed companies and the United States, according to Laurent Gaberell, an intellectual property and agriculture analyst for Public Eye, who has been deeply involved in monitoring compliance with the treaty. Keep an eye out: tensions are likely to rise as our dependence grows on those indigenous seeds, and companies and their allies move to access them with as few payments as they can get away with.

**ALARMED BY THE** corn catastrophe, the National Academy of Sciences decided to figure out how many of America's crops might be similarly vulnerable. For the first time, they set about documenting the genetic building blocks for the seeds of American agriculture. The corn blight was exhibit number one in the cautionary tale, a direct consequence of the corn seeds being "as alike as identical twins." Exhibit number two was drawn from an even more devastating experience a century earlier—the Irish potato famine, which unfolded in ways that were eerily similar to the American corn blight.

In the fall of 1845, in the fields across western Ireland, in acre after acre, the stalks of potatoes turned black, shriveled, and died. The blight spread rapidly across the country. Over the course of about a year, practically the entire country's potato crop never made it out of the ground. The devastation of the potato harvests led to the deaths of at least half a million Irish farmers and their families, and led millions to flee and immigrate to the United States.

The potato blight that reshaped America as well as Ireland was due to the fact that almost all potato plants in Ireland were descended from just one strain of seed brought to the island in 1654 by Sir Walter Raleigh from where they originated, most likely Peru or Colombia. Andean potatoes took well to the moist, foggy climate of Ireland, and within a hundred years that one variety was the primary food source for the nation's peasantry (the royal and merchant classes had access to a far greater selection of foods). Then all it took was a single fungus, *phytophthora*, to spread like a plague onto practically every nearly identical potato in Ireland.

The Academy's report, released in 1972, portrayed the alarmingly thin foundation and genetic vulnerability of the seeds that underlie the American food system.[3] Most major crops, the Academy determined, are "impressively uniform genetically," and thus subject to future attacks much like the one that struck the nation's corn. The corn blight, it said, "is but one example of uniformity," which was threatening to spread throughout US agriculture. It would not be the last such example.

Bill Tracy, head of the USDA's National Maize Germplasm Committee and a professor of Plant Genetics at the University of

Wisconsin-Madison, told me that after the 1970 corn blight, there have been at least two major corn-crop wipe outs ascribed to genetic uniformity. In the eighties, the Goss Wilt bacteria lodged in the leaves across millions of acres of corn fields planted with the identical B-14 hybrid; it took several years to develop resistant varieties because most parent lines were equally susceptible. In the early nineties, a popular hybrid from Pioneer, now owned by DuPont, was found to be vulnerable to the grey leaf spot fungus—which has an affinity for corn residues in hot, humid conditions. The grey leaf epidemic led to yield losses of 10–15 percent across the Midwest,[4] representing billions of dollars in crop losses (many of which were covered by crop insurance payouts supported by US taxpayers). Those hot and humid conditions are precisely the conditions that are occurring more frequently across the Midwest, according to the National Climate Assessment, compiled in the latter days of the Obama administration before the Trump administration's EPA chief shut down the nation's climate research.

In other crops, there have been a number of uniformity-linked crop epidemics—including bananas in Central America and the Philippines, papayas in Hawaii, and on rice mono-crop farms in Africa and Asia. While researching this book, I took a ten-day trip to Jordan to teach a workshop in environmental journalism to journalists from throughout the Middle East, sponsored by the Arab Network for Investigative Journalism (ARIJ). I discovered that several of the participating journalists from Egypt and Tunisia were themselves working on stories about declining diversity in their seed resources as international seed companies moved aggressively into the region.[5] In Jordan and neighboring Palestine, they'd had their own experience of uniformity-induced disaster—except their experience was with tomatoes, a Jordanian plant scientist who works for one of the global seed companies and requested anonymity, explained to me. After the government encouraged the consolidation of farms into tomato monocultures—easier to package and export—dozens of varieties that had been legendary for their juicy, pungent taste were bumped off the market. Then came a double whammy of yellow leaf virus and a surging population of white flies which destroyed as much as 90 percent of the tomato crop. The impact was devastating to the nation's farm economy and food supply. The major difference between the Jordanian tomato and American corn wipeouts was the difference between a vine and a stalk. "We still get a few tomatoes that survive on an infected vine," the scientist told me, "but if the corn plant is infected, you get no corn."

The most vivid example is still unfolding in its multiple tragic consequences: in 2010, Syria was on the fourth year of a record-breaking drought that put unprecedented stresses on the nation's food supply, heightening already high-wire tensions in the country. In the years preceding the drought, the government had been steering crop subsidies to shift the country's agriculture toward crops aimed for export. Local and diverse farms were consolidated in favor of large mono-cropped plantations of primarily wheat and cotton, aimed at markets in Europe, China, and elsewhere. When the drought hit, there were few local varieties left that had the capacity to adapt. Crop wipe-outs followed, food prices skyrocketed, and tensions spiraled. This was one contributing factor to the pressures that have blown apart Syria, a cautionary tale of the frightening and destabilizing influence of diminished seed diversity.[6]

**BACK IN FORT** Collins, Colorado, walking through that bracingly cold seed storage freezer in Colorado in the mid-1980s, I had my first glimpse of what seed diversity looks like. I recall marveling at the colors and shapes of seeds from around the world.

Meanwhile, outside, in the world where most of us non-farmers live, I was getting a sense of what diversity sounds like. New sounds were all over the radio, coming from Latin America, Africa, and Asia, brought to us courtesy of the newly christened genre of "world music." The music rode in on mixing and colliding musical traditions, vibrant and eclectic, exciting hybrids of sound. When I later heard the great American rock guitarist and producer Ry Cooder collaborating with the great multi-instrumentalist musician from Mali, Ali Farka Toure, it felt like I was hearing a powerful reflection in music of the interchanges that occur in the natural world.[7]

Years later when I gave a reading at the Hawaii Book and Music Festival, in a state that has been at the forefront of celebrating and nurturing its rich troves of sub-tropical seeds, I had the good fortune to be preceded by the author of a book on Hawaiian guitar.[8] The author, John W. Troutman, described how American blues musicians picked up a lot of ideas from roving Hawaiian guitarists—who, in the days of Jim Crow laws, were compelled to stay in the same segregated hotels in the South as traveling black blues musicians. There was something in the way that musical ideas flowed easily across boundaries, even artificially created ones, that reminded me again of seeds.

Just as US colonists borrowed from English sea shanties and Irish folk tunes to create "American" country music, and white rock 'n rollers borrowed—or stole—from black American blues singers, seeds are a jumble of influences. They're certainly cosmopolitan, traversing national borders, growing one place and carrying with them the information imbedded in their genes to other places. They're world travelers of the botanical sort, refugees everywhere except the places that are their biological home, a concept which I would come to learn is critical to understanding the great potential, and challenges, of our global pool of seeds.

To get at how the essential principles of exchange and influence work in seeds, let's take a quick journey back in time to a story of two plant breeders, a Russian and an American, who in the 1900s introduced us to the importance of genetic diversity. Of course, farmers in every culture have known for millennia why multiple varieties are important for a field's vitality. But in the Western world, it was these two men who put the science into plant breeding, and showed us the extraordinary possibilities of an agriculture rich with genetic diversity, and the implications of losing it.

**I CAME UPON** Nikolay Vavilov, a legendary Russian botanist, when trying to figure out why places like the seed storage lab in Colorado matter. Several plant scientists and explorers I encountered along my reporting journey told me that he was a primary influence on their work. So, I did what anybody would do back when, imagine, there was no Internet; I went to the library. I sat down at a long brown wood table at the main branch of the San Francisco Public Library and, in that sacred murmuring silence, briskly whisked through card catalog drawers searching "seeds," "Russian botanist," and "Vavilov."

There was little to be found in the non-academic press, but there were articles in scientific journals about Vavilov's extraordinary journeys to every corner of the earth in search of seeds to strengthen the ability of crops to withstand Russia's harsh conditions. It's been a hundred years since, but there's no question that our understanding of the central role played by seeds to a vibrant and resilient agriculture owes a lot to Vavilov's journeys traversing the globe in search of seeds to mix in with the Russian gene pool. In St. Petersburg, he began what would become, for a time, the largest selection of seeds in the world, including new varieties of wheat, rye, barley, oats, and other cereals that could survive Russian winters.[9] His staff would later famously defend those seeds against the invading Nazis during World War II.

Nikolay Ivanovich Vavilov grew up in Moscow, where his father was a textile merchant. His family was part of a small and highly educated mercantile class at a time of great creative and political ferment in Russia. At Moscow University, he studied plant pathology—how plants die, which is really a way to study how to keep them alive.

In 1916, with a fresh PhD, he was dispensed by the czar to figure out why Russian soldiers at the country's southern border with Persia were getting sick from eating local wheat varieties. He discovered that they were inadvertently consuming a fungus on the wheat; that fungus later became quite familiar to contemporary scientists as the progenitor of lysergic acid, a.k.a. the psychedelic drug LSD. The troops stopped hallucinating when he convinced the commanders to change their source of wheat.

From there, Vavilov headed into the Pamir mountains of Central Asia to test a supposition: that the wild relatives of wheat, rye, bulgur, corn, and other cereals and grains could be bred into Russia's seed-stock to strengthen their ability to survive the country's harsh winters and hot summers. He sought seeds that had resisted and survived the inclement conditions, catalogued them and over many years sent bags full of them back to St. Petersburg. In the process, Vavilov made visible the genetic ink with which the stories of our planet's seeds are written.

When the Russian revolution swept the czar out of power in 1917, the new Communist government did not sweep Vavilov out of his position as they did many other university-trained career scientists and businessmen. In a sign of Vavilov's already luminous stature, the new premier, Vladimir Lenin, commissioned him to expand upon the quest he had begun under the czar, and, in 1921, appointed him director of the St. Petersburg Institute of Plant Industry.

The Russian plant explorer went on to traverse the globe and was the first botanist to achieve international acclaim for the audacity and breadth of his forays to distant and often forbidding locales in search of seeds to bring back to St. Petersburg. Photos of Vavilov from the 1920s show a dapper and handsome man with a lustrous mustache, in a finely cut suit with a narrow and tightly cinched tie, and a hat deftly angled on his head. He was by all accounts an impressive conversationalist—you'd have to be to gain the confidence first of Czar Nicholas II and then Vladimir Lenin.

Vavilov went off for months at a time clambering through valleys and peaks, over rugged mountainsides and into lush river basins, to collect samples from farmers who had figured out over centuries how to grow food in volatile weather, with wild seasonal swings from cold to heat. He observed on every inhabited continent that the damage inflicted by diseases or pests was significantly reduced when farms contained a mix of locally-adapted varieties.

Over some forty very active years, Vavilov organized more than a hundred expeditions to sixty-four countries in search of wild and domesticated plants and their germplasm. On his seed quests, Vavilov traversed multiple frontiers—not only between nations, but between peoples. Many of those he encountered were indigenous people, in the horn of Africa, the mountains of Central Asia, the Amazon jungle in Brazil, and in the southwest of the United States—people who had been cultivating crops adapted to their ecological niches for centuries. A master of multiple languages, including English, he met with tribal leaders and diplomats, with scientists and shamans. One key thing Vavilov learned on his outings, which he recorded in his journals, was that farmers' own local knowledge—too often discounted as "non-scientific"—offered important insights into how to cultivate crops able to withstand environmental stresses. Most significantly for our present-day purposes, among the places Vavilov explored were some of the planet's driest and hottest regions.[10]

Vavilov's far-flung forays led him to discover a hugely important truth about the earth's ecology: 90 percent of the bio-diversity of our planet's plants and animals originate from a narrow band that circles the equator. He understood that this equatorial belt was key to the genesis and evolution of new species and critical to the world's food supply. Vavilov was the first to identify "centers of origin" for most of the world's crops. This is why, when disease or inclement weather strikes a crop or region, scientists travel many hundreds or thousands of miles away from home, often to distant mountains, river basins, or coastal wetlands, to find and bring home a plant's progenitors. It's often what are known as 'wild relatives' of domesticated crops (like the teosinte that helped save America's corn crop) that offer the most valuable germplasm, conferring traits of resistance to diseases and pests, and drought in hot climes, that have been lost through the breeding process.[11] Centers of origin for our major food crops include Peru for potatoes, Ecuador for bananas, Indonesia for beans, and the Philippines for rice. If North Americans, who rely on foreign crops for 90 percent of their diet,[12] were to eat only foods native to their continent, dining tables would be pretty sparse—such a nativist menu would include little more than pecans, cranberries, walnuts, artichokes, and pumpkins. In recognition of his pioneering work, the species-rich regions running through the equatorial middle of the planet are now frequently referred to as "Vavilov Centers."

* * *

**WHILE VAVILOV TRAVERSED** the world in search of seeds, another plant breeder, an American, had a different strategy for seed collecting. Luther Burbank barely left his home in Sonoma County, California, where he became the world's first—and possibly only—celebrity breeder by bringing the seeds of the world to him and testing them out in the loamy fertile soils of Northern California.

Burbank was born in 1849, thirty-seven years before Vavilov, and grew up in the town of Lancaster in western Massachusetts. At the age of eighteen, he discovered a potato plant in his family's garden that was larger, contained more starch, and did not rot as quickly as the other potatoes he'd planted. He sold the seeds to a local nursery, which turned them into one of the most popular potato varieties of the time. Some were even sent off to Ireland, still struggling to expand its breeding stock from the limited gene pool that had led to the famine. A derivative of Burbank's original potato, the "Burbank russet," remains popular today.

Burbank used proceeds from the sale to finance a trip out west, where he discovered the fertile farmland in Sonoma county, and stayed. He planted a tree-fruit orchard and experimental garden in the town of Sebastopol, about fifteen miles from his home in Santa Rosa. There, he grafted botanical cousins onto one another to introduce hardier and tastier varieties of apricots, peaches, nectarines, and plums, and bred dozens of new varieties of beets, potatoes, squash, rhubarb, tomatoes, peas, peppers, corn, and much more.[13] (Burbank's home and surrounding garden is now a museum in Santa Rosa; a portion of his orchard remains standing, along with an exhibit, in Sebastopol.)

He was tall and gangly, with a long Yankee face. A homebody and somewhat misanthropic—he seemed to prefer the company of plants to people –Burbank left California just once over his five decades in

Sonoma. But he understood that an abundant variety of seeds was key to successful plant breeding, and used the US mail to get them. He ordered seeds from breeders throughout the American South and New England, from Native American communities in California and Arizona, and even from breeders as far away as Japan, Mexico, and Australia. His graft of apricot and plum gave us the pluot, one of the most popular summer fruits on the market today. When I visited the Burbank Experimental Farm and Orchard on the outskirts of the town of Sebastopol, the kindly docent who showed me around stopped under a tree sagging with plums. The tree, she said, was the product of plum seeds and grafts sent to Burbank by a Japanese breeder in Yokohama, which Burbank crossed with plums grown by native tribes in and around Sonoma. The result: plump and juicy Santa Rosa Plums, which still drop like baseballs from that tree when they're ripe, and the descendants of which are still widely available in fruit markets across the United States. Burbank apparently had friendly relations with native tribespeople, and offered them samples from his new creations, but appears to have offered little more than minimal purchase fees as financial compensation for the invaluable seed resources they provided him. It would take the global treaty, some sixty years later, to even begin the process of officially acknowledging the debt we all owe to those indigenous cultivators.

Burbank also availed himself of the abundant quantity of free seeds disseminated at the time by the Department of Agriculture, which was encouraging agriculture west of the Mississippi. That practice of free seed distribution was curtailed in the 1920s under pressure from the American Seed Trade Association, the trade group for a nascent seed industry. Burbank launched the United States' first national-scale seed catalog in 1893. A catalog of Burbank's botanical contributions runs for 101 pages.[14]

The two men met just once, in the autumn of 1921, when Vavilov visited Burbank at his home in Santa Rosa after attending a scientific conference in San Francisco. Their personalities were just about as different as the countries in which they flourished. Vavilov was gregarious, Burbank an introvert. But the American and the Soviet Russian crossed the considerable barriers of personality and politics to connect over plants. Their meeting occurred five years after the Russian Revolution. The United States had spent millions of dollars trying to prevent the Bolsheviks from taking power. Now Russia's most famous scientist would meet America's most renowned seed breeder. The desire to share insights about breeding more resilient and diverse seeds would trump the lingering tensions between the capitalist country and the newly communist nation.[15]

At the time, Luther Burbank was botany's equivalent of a rock star, celebrated for having bred new varieties of fruits, nuts, and flowers in his orchard and on his experimental farm. *TIME* magazine called him the "plant wizard." Vavilov was revered in scientific circles, speaking at the world's most distinguished gatherings—from Oxford to Harvard to the University of Tucson, in Arizona.

Although I could find no record, alas, for what transpired between the two scientists that day—other than various records indicating that the meeting occurred—we know that they each had a major influence on the other's thinking, bridging the very substantial gaps between Russia and the United States through their common desire for a healthy and vital food system. We know that shortly after his meeting with Burbank, Vavilov established an outpost in the United States—the Department of Applied Botany and Breeding, based in downtown Manhattan. The enterprise, which had the blessing of the USDA, employed Russian and US scientists to research what could be learned from North American cereals and grains for possible introduction back in Russia. This brief botanical détente lasted from 1921 to 1925.[16]

We also know that they read and appreciated each other's work and were aware that they were the front-line pioneers in applying Charles Darwin's insights on evolution to agriculture. Although plant breeders had over millennia practiced Darwinian principles, they did not have scientific foundation to fully understand how they were using them. The principle of survival of the fittest, after all, wasn't invented by Darwin, but was identified and explained by him. The monk Gregor Mendel may have figured out in the 1860s how to breed and obtain desired characteristics through meticulous mixing and selecting of different parent lines, but it was Darwin who showed us *why* Mendel's famous genetic progressions with a plot of sweet peas actually worked. And it was Vavilov and Burbank who showed us how to use that knowledge, and apply it with multiple sources to obtain evermore robust varieties.

Burbank's enthusiasm for Vavilov's numerous botanical discoveries helped burnish the Russian botanist's reputation in the United States, and thus contributed to disseminating his ideas far beyond Russia. Similarly, Vavilov's acknowledgment of Burbank's ingenuity with plants would greatly enhance the reputation in scientific circles of the self-taught American, who had never achieved a college degree. Upon Burbank's death in 1926, Vavilov wrote a glowing obituary for a Russian newspaper, an effusive homage to the originality of the American's work in plant genetics: "The uniqueness of Burbank's garden is that everything in it that meets the eye is the result of creativity. Everything in the garden has been subjected to the influence of the plant breeder . . . an entire living museum in which everything was full of meaning."[17]

Together, Nikolay Vavilov and Luther Burbank represent key turning points in the evolution of modern agriculture. Their insights are important parts of the root system upon which farmers, scientists, and activists have built a deeper understanding of what is necessary to save our food in the face of climatic turmoil. They experimented with new combinations, had the patience to await the results, and the knowledge to understand the significance of what they were seeing. When they unlocked the combination to the genetic safe, treasures spilled forth.

But what they could not have imagined was that the genetic secrets they'd unlocked could also be re-locked, the knowledge they hoped to spread could be restricted, and the diversity that they'd hoped to expand undermined. Once genetic resources were identified and understood, they could also be reined in.

THE NATIONAL ACADEMY of Sciences report on the 1970 corn blight was scathing in its assessment of who was responsible for the mounting dangers to our food system. It wasn't due to the usual ebb and flow of good years and bad years, to the plant vulnerabilities that have been with us since the beginning of agriculture. Rather, responsibility lay with the actions of humans, specifically those with authority in Congress who had been steering us toward evermore monochromatic agriculture with subsidies, insurance programs, and tax incentives. "This uniformity," they concluded, "derives from powerful economic and legislative forces."[18] They pointed toward a perverse equation: the larger the farm, the less variety in the seeds cultivated into food. What happened to Fred McLain and thousands of other corn farmers, they warned, could happen again.

By the time the Academy issued its final report, Vavilov had been dead for thirty years. He was ultimately, tragically, trapped by his own enthusiastic collaboration with Western scientists. Accused of "bourgeois" sympathies, he was arrested by Stalin's policemen twenty years after his meeting with Luther Burbank and died in prison nine months later.[19] Struggling financially over his lifetime, Luther Burbank advocated for a patent law to protect his botanical creations. A version of the law that Burbank advocated in his later years, the Plant Patent Act, was passed in 1930, four years after his death, and signed by President Herbert Hoover. It permitted the patenting of asexually created plants—new varieties created with grafts and cuttings, which is how Burbank did most of his work. It came nowhere near the expanded level of patent protections for sexually reproducing plants instituted by Congress some fifty years later, which would open the door to the seed industry we see today.

What neither man could have possibly foreseen was the extent to which, over the decades to come, the seeds and the knowledge they had shared with the world and with their fellow scientists would be deemed some of the world's most valuable intellectual property and locked behind patent laws. Neither could they imagine the idea of obtaining genetic material from an unrelated species—a salmon for example—and inserting it into the DNA of corn, soybeans, or strawberries in order to create a single desired trait.

Nor could they have had the slightest inkling of the challenges we'd face almost a century later from climate change. These were the early days of internal combustion engines and no one was thinking of the collateral damage these transformative inventions would have on the balance between water, sun, and soil that is the basis for growing food.

The two may have been long gone, but they might as well have ghost-written the Academy's recommendations. Among the most notable: to avoid another corn blight or similar crop epidemics, the United States should broaden the gene pool for all crops, encourage the broad dissemination of different seed varieties, and preserve native and wild varieties as source pools of future variation. These measures, the Academy said, were the tools we needed to avoid a repetition of what happened to Fred McLain and thousands of other corn farmers. It would give agriculture more resilience to deal with changing environmental conditions, diseases and pests, and a more deeply rooted food security.

Since then, practically every major step taken by the United States has sent the country in the opposite direction. Step by step, more and more seeds are being pushed off the land where they're intended to grow and, as a last resort, into the cold vaults tucked into the mountains of Norway or the campus of Colorado State.

The new players have new names. Scientists like Burbank and Vavilov—and there were others, too, like W. Atlee Burpee, George Washington Carver, and Henry Wallace, founder of Pioneer Seeds and vice president during Franklin D. Roosevelt's third presidential term—have now been succeeded by company names like Monsanto, Syngenta, and DuPont.

The Academy of Science's report was intended as a warning. But it now reads more like a chronicle of the next half-century, as Congress and the courts pushed the United States toward an ever-greater concentration of the seed industry and ever-more precarious genetic base for our food.

# CHAPTER 3

# SEEDS, INC.

·

**I FOLLOWED THE** trails of genetic information embedded within the seeds in the fields of America, and ended up in a leafy and carefully manicured cul-de-sac in the suburbs outside of Alexandria, Virginia. Here is the six-story steel and glass repository of America's intellectual property, the United States Patent and Trademark Office. At the far end of the cul-de-sac is an exhibit harkening back to the nation's earliest patents. There I saw the nation's very first, a patent for making potash, an ingredient in fertilizer—framed, on the wall, and signed in 1790 by George Washington.

Two hundred and twenty-six years and more than six million total US patents later, I paid them a visit. A veteran patent examiner (who requested I not use their name) showed me how concentration in the seed industry was occurring before our very eyes—that is, if your eyes happen to be focused on the Patent and Trademark Office's list of who's getting patents on which plants. You can see the transformation from local to global rendered in the tiny print. Starting in the mid-1990s through 2010, the number of patents assigned to distinct individuals or small regional seed companies with unfamiliar names steadily dropped, and those assigned to corporate enterprises with highly familiar names (Monsanto, DuPont, Dow, or their subsidiaries) steadily rose.[1] Over the course of two decades, power over our seeds shifted from a loose network of regional seed companies, close to the ground on which farmers till their fields, to distant national or international companies, connected to the imperatives of stockholders and to the vagaries of chemical production, and ever-further from the fields where they are planted.

The seed oligopoly we see today was built upon the conversion of freely grown and exchanged seeds into seeds as intellectual property. This was accomplished largely through a sequence of historic legal decisions on the other side of the Potomac River at the Supreme Court in Washington, DC.

First, there was the matter of extending patent protections into the realm of living organisms. General Electric, the giant industrial and consumer products company, led the way. One of GE's scientists, Ananda Chakrabarty, developed a bacterium that degrades oil, handy for the oil industry to clean up oil spills. The company's first patent application, in 1980, was denied the US Patent Office. Previously, patents had been granted for the *process* utilized by living organisms—say, yeast fermentation. But GE wanted a patent on the bacteria itself, a frontier the Patent Office was unwilling to cross, on the grounds that something alive and capable of evolving could not be patented. The company appealed the rejection. The case went all the way to the Supreme Court, which decided in the company's favor. There was little difference, the Court declared, between patenting a living organism, like bacteria, and an invented object: ". . . [T]he fact that micro-organisms are alive is without legal significance for purposes of the patent law."

It would take some years for the principle affirmed by Chakrabarty to ripple decisively beyond the realm of bacteria.[2] But when it did the seed industry would never be the same.

A relatively small St. Louis-based chemical company called Monsanto, known mostly at the time for producing a range of highly toxic chemical herbicides—including Glyphosate and the defoliant Agent Orange, used widely by US forces during the Vietnam War—saw the writing on the wall. In 1982, Monsanto acquired its first seed company, the Arkansas-based Jacob Hartz Seeds, which specialized in soybeans.

Then came a new chapter in the transformation of seeds from evolving organisms into patented property. In 1985, the Patent Office's Board of Appeals approved a bio-tech company's request to patent the biochemical mechanisms by which a plant takes on desired characteristics. The company, Molecular Genetics, was granted a patent for its DNA sequences and the mechanisms by which a seed expresses characteristics, say, resistance to heat or frost. In other words, the processes for creating specific traits, not just the seed itself, could be patented. These were known as utility patents, and they'd help turn seeds from a business backwater to multi-billion dollar profit centers.

In 1995, the Supreme Court pushed the ability to enforce the new plant patents a step further: an Iowa farm couple, Dennis and Becky Winterboer, was accused by a seed company, Asgrow, of reselling its patented soybean seeds. The Supreme Court affirmed Asgrow's claim that this violated the company's patent rights, and defined those rights according to traditional contract law. Seeds were characterized not as singular evolving organisms, but as bundles of "traits," to which farmers gain access through patented seeds. "Traits" in seeds could be patented and branded like other products.

Previous plant patents, like those that Luther Burbank advocated, had covered only asexually reproduced plants, in which breeders used grafts or cuttings to create new varieties. Now, you could patent plants capable of mating with other plants, combining genes that may very well have preceded the breeder's interdiction in the evolutionary process. Take a seed embedded with thousands of years of evolutionary changes in its DNA, add or change a trait, and the result can be patented, as if a breeder had invented the whole plant, notwithstanding the backstory to its genes. Thereby a chain is broken that once connected today's farmers to their predecessors, as planters, cultivators, and breeders of seeds grown in their fields. Found varieties can also be patented, so that if a new variety is "discovered" sprouting in a farm or garden or research center—technically "found" seeds must be discovered in an already cultivated area—a patent can be obtained on a plant the breeder had no role in creating.

"Nature does 90 percent of the work in selecting adaptations, indigenous farmers have done 8 percent of the work, and modern plant breeders maybe 2 percent of the work," commented Gary Nabhan, an ethno-botanist who has worked for decades with plant varieties in the desert. "And the big companies they work for claim the patents."

In this regard, the patent system in Europe departs significantly from that in the United States. When, in 2015, Monsanto obtained a patent for a melon with pest resistance that is based on a melon seed in wide and common use in India, sixty-five thousand European citizens signed a petition circulated by a coalition of NGOs, including the German group No Patents on Life and the Swiss group Public Eye, calling on the European Patent Office to repeal the patent. In May 2016, they succeeded: the EPO took the groundbreaking move of repealing Monsanto's patent.

Similarly, when the Swiss firm Syngenta was issued a patent on a pepper plant that is resistant to two major pests—trips and whiteflies—the Swiss NGO Public Eye launched an investigation and revealed that the seed actually had its genesis in the pepper fields of

Jamaica, and had been stored in a public seed bank in the Netherlands. The group and a coalition of European NGOs led a public pressure and lobbying campaign that succeeded in convincing the European Parliament to request that the EPO stop offering patent protection to already publicly accessible seeds. "We are trying to stop them from privatizing public resources or the public resources inside of the gene banks," explained Laurent Gaberell, a seed policy expert with Public Eye, based in the Swiss city of Berne. In June 2017, the EPO agreed to reform its patent system and, in a divergence from the US approach, no longer permits the patenting of seeds that could evolve from natural processes—like those which gave rise to the Jamaican pepper plant.

**BUT WE'RE GETTING** ahead of ourselves. Back on the US side of the Atlantic, US patent laws were still being defined, and the transformations they triggered taking shape. A year after the decision favoring Asgrow in 1995, that company was acquired by Monsanto for what was then a record-breaking $240 million. It was off to the races for corporate consolidation. Chemical companies in the United States and in Europe (where the main players are the German chemical firms BASF and Bayer, and the Swiss chemical firm Syngenta) soon followed Monsanto's lead and went on a seed buying spree.

"Monsanto and other companies recognized how much more money they could make by selling chemicals and seeds," said Philip Howard, an associate professor at Michigan State University who has been documenting agricultural industry consolidation over the past several decades. Howard's book, *Concentration and Power in the Food System: Who Controls What We Eat*, tracks the steady narrowing of players across agriculture, including seeds, like a shrinking family tree. "They got into the seed business to sell more chemicals, and the others soon followed," Howard concluded.

Lock in the seed, and you can sell the chemical to go with it—two sides of the same coin. With a twist: the company sells its product on either side of the coin. Seeds can be bred and engineered to grow in association with chemicals that the mother company also happens to produce.

By the end of the twentieth century you didn't need a farmer's almanac to know which way the wind in the seed business was blowing. Between 1980 and 2000, more than a thousand seed companies were bought by petrochemical, pharmaceutical, and commodity grain companies, most of them snapped up in the wake of the legal decisions strengthening patent rights. Monsanto alone sells its seeds not only under its own brand name, but under the brand names of more than fifty formerly independent companies, including Asgrow, Seminis, and Hartz.[3] DuPont, the chemical giant, went on a buying spree—including the legendary seed company Pioneer, founded by FDR's former Vice President Henry Wallace. By 2017, DuPont had merged with Dow. Hundreds of seed varieties, bred by formerly independent companies to be responsive to local conditions, were replaced by blockbuster seeds to be grown across vastly different terrains, as long as they were boosted by chemical inputs.[4]

Patent protections were and continue to be aggressively enforced. Legal attacks were unleashed on farmers accused of illegally planting or selling patented seeds. The most infamous of those cases involved the Canadian farmer Percy Schmeiser, who, in 1997, discovered several dozen of Monsanto's patented Roundup resistant canola plants in his field in Saskatchewan. When he replanted some of that seed on his farm the following year, Monsanto representatives demanded that he pay them a licensing fee. Schmeiser refused, claiming that the original seeds had traveled to his field without his knowledge. Monsanto sued for patent infringement in a case that was ultimately decided in the company's favor by the Canadian Supreme Court.

Altogether, Monsanto reports that it has filed 147 patent violation complaints in US courts against farmers on similar grounds since 1997, and they won all nine cases that went to trial.[5] In other cases, the company reached settlements with farmers, and did not disclose the terms. The expensive and intrusive cases sent a powerful signal—patented seeds were no longer fair game for the usual exchanges and experiments that had for centuries been part of life on a farm.

Today, many farmers are, legally speaking, no longer *purchasing* the seed that they plant but, rather, are *renting* the use of the traits that the seed contains. And their right to use those traits expires at the end of the season. Which seems like a perversely perfect business solution to a perverse market problem—the steady drop in the number of farms and farmers. Federal subsidy and insurance policies combined with the high financial risks of farming and a steady downward pressure on prices from commodity trading companies have led to a steady drop in the number of farms since the 1980s. Between 2007 and 2015 alone, according to the USDA, farms declined from 2.2 million to 2.07 million, and that number is expected to continue dropping. Companies can sell the same patented seeds to the same farmers year after year, along with the chemicals that go along with them.

As I sat at my desk trying to parse out the vaporous abstractions of intellectual property law applied to plants, I took a break. I switched my Pandora stream from jazz—Brubeck and his musical cousins, good for quiet writing—to blues—Buddy Guy, Muddy Waters, and their musical offspring, which got me out of my chair, pacing (an improvement on smoking). And then it hit me: the concept behind seed patenting as it's been shaped by the courts is not that different from how I was listening to my music. With music, you pay for access to the technology which delivers tunes, but you don't ever "own" the music. You consume it. Just like farmers and seeds. And what about the little box you check for all those new computer applications, which state that a "click" constitutes your agreement to abide by copyright or trademark restrictions? It's similar to the "rip-tag" attached now to most bags of seed, with tiny type affirming that the act of ripping the bag open is the equivalent of signing your name to a contractual agreement to abide by patent restrictions, which usually prohibit replanting or resale of the seeds.[6]

The key difference between food and music, of course, is that we music consumers are not making the tunes. Unless we are budding DJs bootlegging samples, we passively consume it. Farmers, needless to say, grow our food, and yet are being distanced from the primary ingredient for doing so—an ingredient they have had a rightful claim to since the earliest days of agriculture.

At just the moment when the rising heat, diminishing water, and expanding array of diseases and pests require new combinations of genetic codes, our palette of seed options is narrowing.

CONCERNS OVER THE tightening of control over farm industries prompted an unprecedented effort by the USDA and Department of Justice to join forces in 2011 to assess the impact of mergers and acquisitions in four major agricultural sectors, including seeds. Top officials from each agency hit the road together and held a series of public hearings. Farmers testified about their fears over the impacts of genetic uniformity and about being stove-piped into buying, and thus planting, from a dwindling number of seed options—bad for business, they said, and bad for their ability to sustain resilient farms.

Then nothing happened. The effort ended with a whimper: a report was issued sounding alarms over the constraining impacts of consolidation, and declared the inadequacy of anti-trust law to do anything about it.[7]

Five years later, in September 2016, the farmers tried again. Roger Johnson, president of the National Farmers Union (NFU), representing more than 200,000 independent farmers, led about 250 of his members to Washington, DC to lobby their congressional representatives and Agriculture Secretary Tom Vilsack to slow down the spate of consolidations. After a flurry of DC meetings, the farmers went home. Again, nothing happened.

But in the industry, a lot happened. Since the collapse of the USDA-DOJ anti-trust inquiries in 2012, the seed industry underwent a paroxysm of acquisitions and mergers. By 2018, we were left with the trifecta at the top: Dow and DuPont merged into one, the world's biggest single chemical company and third biggest seed company; ChemChina, owned by the government of China and now the world's second biggest chemical company, purchased the Swiss firm Syngenta, the world's fourth largest seed company; and Bayer, Germany's biggest chemical and pharmaceutical company, acquired Monsanto, a combination that creates the world's largest seed and chemical company.

Back in the 1980s, I coauthored a book, *Circle of Poison* which revealed how dangerous pesticides banned in the United States were exported overseas and returned to Americans as toxic residues in food.[8] At that time, the chemical companies were still distinct from the seed companies. Now the producers of the majority of the world's commercial seed varieties are one with the producers of the chemicals.

"After all these mergers," commented Roger Johnson, head of the farmers union, "what we're seeing is fewer seed choices at just the time that climate change is making everything more variable." He said the effect has not been what the companies told regulators—that there'd be more innovation—but rising prices and diminishing options.

A secret history of the seeds in widespread use across the Midwest was revealed by the Farmers Business Network in 2017. The FBN was founded the previous year by a group of Silicon Valley engineers, many of them with farming backgrounds. They use sophisticated data collection technologies to provide information for farmers that is independent of the seed and chemical company-funded sources on everything from the weather to commodity price trends to comparative pesticide, fertilizer, and seed performance and pricing.

With an eye on expanding into seeds, the company took a very Silicon Valley-inspired approach: they crowd-sourced a survey of their farmer-members' seeds, requesting seed samples and the bagtags that go with them.

The response was overwhelming, far greater than they anticipated, Charles Baron, a co-founder and senior vice president of the company, told me. In came more than seven thousand samples within a month. Over the summer of 2017, the company conducted genetic analyses of the samples, and, by the fall of 2017, came to a startling conclusion. Many of the samples were actually the same seeds, though they came in different packaging: 38 percent of the soybean seeds and 45 percent of the corn seeds overlapped precisely with the seeds of at least one other company.[9] Many of the seeds sold by Monsanto, Dow-DuPont, Syngenta, and Bayer, and several other smaller companies, are the same. Baron commented: "Farmers want to build up resilience with multiple seed varieties so their field is not susceptible to the same disease or pest. But we discovered that when farmers think they're buying a mix of seeds, they're often buying duplicates with different labels." Farmers who think they're protecting themselves from a multiplicity of threats by diversifying their seeds are frequently just planting the same seed over and over. Consolidation has not led to exciting new seed varieties in America's major crops, but to the same seeds in different packages.

When I visited the Fort Collins Seed Storage Lab back in the 1980s, the United Nation's Food and Agriculture Organization (FAO) declared that at rates of extinction common at the time, three quarters of the vegetable seed varieties planted in Europe could be extinct by the end of the twentieth century. In retrospect, the FAO's warning seems sadly understated. By 2015, a quarter century later, the FAO declared that three quarters of *all* the world's crop varieties that were around at the turn of the nineteenth century had become extinct.[10]

We can see that decline across the United States. Researchers at North Dakota State University and Kansas State University conducted a detailed look at the seeds planted in every agricultural county between 1978 and 2012, and discovered that the range of cultivated seeds over those thirty-four years shrank in almost every region of the country. (The sole exception from the trend toward genetic uniformity was in the Mississippi Delta region, which is showing increasing diversity as cotton cultivation drops and is replaced by food crops). The result, they predict, are yield declines as climate disruption accelerates: "[O]ne important consequence of increased crop homogeneity is the potential for yield instability with anticipated increased unpredictability in weather patterns associated with climate change."[11]

Not only are local seed varieties being displaced by vast fields of identical seeds, the future adaptations that the lost parent lines could create if given the chance to mix it up in the gene pool are forever eliminated. This is occurring just as growing conditions are reaching unprecedented levels of uncertainty.

\* \* \*

**TO UNDERSTAND MORE** about what this convergence of consolidation, climate disruption, and seed extinctions means for our food crops, there's probably no better group of people to talk to than the National Association of Plant Breeders—which happened to be having their national convention in the summer of 2017 in Davis, California. I caught an Amtrak train in Oakland and an hour later arrived at the college town, home to one of the leading universities for agricultural research. Over a sweltering couple of August days, the men and women who devote their careers to devising the next generation of crops gathered to compare notes from the field, and to share their sense of alarm about decreasing seed diversity and increasing environmental disruptions.

It was hot and muggy outside, like a science experiment on tolerance for humidity. But inside, the air-conditioned hallways were lined with photos of pests and fungi that follow the heat and freakish weather to attack crops in a spectacularly morbid number of ways— fungi sprouting in soil left moist after a downpour during a heatwave; fly larvae once killed by the onset of cold winters now able to hatch on wheat leaves when the cold weather comes later in the season; new pests with grotesque proboscises that once happily proliferated in Mexico and Central America making their way north into the warming American southwest.

Other displays showed multi-colored charts documenting the shrinking budgets of public sector seed breeding and research programs. The two visual threads seemed somehow, jarringly, connected. Research by public institutions, pursuing research focused on devising more resilient and nutritious crop varieties for the public have been the source of American agricultural innovation for more than a century. Land grant universities, ag extension services, and the like have been central players assisting farmers to adapt to changing conditions. And now, as those conditions undergo changes on a level never seen before and new threats emerge, the charts told of a steady decline in funding for public breeding programs, and a steady shift of research and development (R&D) from the public to the private sector.

One such chart, from the University of Wisconsin, Madison, showed a downward plunge from 2000 to 2013 of the USDA's funding of public breeding, from being responsible for roughly 50 percent of all new agricultural R&D to 30 percent. Seventy percent of new cultivars now come from private industry. More and more germplasm—and the R&D that lies behind it—is getting locked behind company patents and trade secrets.

The shrinking number of possible combinations for public breeders to work with is putting our food system on an ever-more narrow foundation, according to Jack Kloppenberg, a professor of environmental sociology at the University of Wisconsin-Madison.[12] Kloppenberg has conducted extensive research on the impacts of plant patenting. He commented: "If you're a public breeder, say, you're very interested in heat tolerance. One of the first things you need to do before you get started is to figure out if anyone has patents on that heat tolerant trait. Do you think that's easy? It's expensive and time consuming. So what public breeders do is stick with the stuff they know—limiting the material they're looking at and working with."

Needless to say, heat tolerance is not an obscure or only occasionally necessary trait: the US National Climate Assessment estimates that the temperature in the United States has risen about 1.5 degrees Fahrenheit from 1990 to 2015, about twice the rate of increase as the previous sixty years, and under current emission scenarios will continue to rise at an accelerated pace. We're going to need as many heat tolerant seeds as we can find.

These are not just numbers on a ledger—the mission of research and development at public universities is to sustain innovations for the public good. The goal of private companies supplanting them in agricultural R&D is, first, to ensure a profitable bottom line for the mother company, which may not be the same thing as ensuring healthy and resilient food. (History is filled with examples in which that has not been the case, from the application of massive quantities of toxic chemicals that are left as residues on our food to modern grains that may be triggering auto-immune conditions among the people who consume them.) The shift from public to private breeding is making us more reliant on a narrower pool of genetic lines, concludes the Union of Concerned Scientists, threatening "our food security. Low genetic diversity in farmers' fields makes crops increasingly susceptible to disease-causing agents, which could spread more quickly and widely than among a more genetically diverse crop."[13]

Breeders tend to talk about evolution as an almost mythically powerful force, with great respect for its omniscient inexorability. "Evolution with human assistance" is how Charles Brummer, professor of Plant Biology at UC Davis, and chief organizer of the NAPB conference, describes his profession. Plant breeders, he said, shape evolution by selecting plants based on their response to pressure and threats—whether from pests, extreme temperatures, not enough or too much water, for a few examples. Do it again and again until a crop is strong and predictable enough for commercial planting.

Brummer is also the director of the UC Davis Plant Breeding Center, engaged with pioneering research on crop adaptations. In their search for new and improved characteristics, they've experienced option after option being taken off the table. "The trend toward tying up germplasm is not desirable," he told me. "The restrictions impede our ability to contend with things like climate change."

What concerns him is the shift in priorities when increasing amounts of the research is in the hands of private companies. What breeders look for, Brummer said, determines what you'll find: "When it comes to plant breeding, you almost always get what you are selecting for. If you are not selecting for it, you won't get it." Translation: if you're not looking for pest or weed resistance because there's a chemical that can accomplish the same thing, then you won't find it.

Not surprisingly, pesticide and herbicide sales continue to skyrocket, reaching at least $10 billion annually since 2000, according to the USDA's Economic Research Service.[14] "Instead of trying to develop inherent resistance of crops to pests, it's more profitable to take a sledgehammer to the system with chemicals," commented Doug Gurian-Sherman, policy director at the Center for Food Safety, an NGO based in Washington, DC that has been at the forefront of calling for agricultural reforms.

As Salvatore Ceccarelli, a plant scientist who works with the United Nations and international food agencies to maintain seed diversity in the Middle East and Northern Africa, wryly put it, "Why should they sell seeds that do not need pesticides? That would be like shooting themselves in the foot."

**WHEN I NEXT** caught up with the McLain family in Iowa, twenty years after my first visit, Frank McLain was running the farm he'd inherited from his parents. He's the fifth McLain to run the place since it was homesteaded in 1862. Frank was fifty, genial and easy to make laugh. It was a sweltering hot day in July. He told me what it felt like to take on the mantle of the farm from his father: "What they passed on to me is the feeling that this land is not just a hunk of dirt that you use and sell," he said. "That a piece of ground is something that should be kept for the next generation, that you're just a steward and you're not just to use it as a tool or as a doormat."

He took me on a ride in his pickup truck on a dirt road through his fields. To the left, we drove past those same fields I'd seen earlier, except now they were planted with the latest innovation to come out of the private breeding centers. Row upon row of corn plants of identical height, all of them about five feet high and containing a genetically engineered trait, the Bt, that makes the plant poisonous to a common corn pest. To the right was a field, stretching seemingly into the horizon, of genetically engineered Roundup Ready soybeans, also mature and green and, from the pickup's window anyway, looking pretty much identical. Frank was the first in the family to plant genetically engineered crops. He even had a test plot of corn nearby for Monsanto, testing out newly engineered seeds before their commercial release—for which he received a premium every year, a little extra, he told me, that offered some measure of protection from the volatile fortunes of farming. Frank said that planting GMO (genetically modified organism) crops had reduced his risks of losses from weeds and the Bt pest, and required far less labor than the crops he used to grow because they don't have to be monitored as frequently. It was his second year growing them, and he was appreciative of how they shortened the workday, reduced his exposure to chemicals (at least with the corn), and provided a good yield for a growing market. They reduced the uncertainties in a profession that is riddled with them. They were new and they seemed to work.[15]

Indeed, this is how GMO's came to most American farmers—as a highly touted new crop innovation, reducing farmers need to till because Roundup did the work of killing weeds; and reducing the need to spray pesticides on corn because every Bt corn plant was engineered to have a toxin that kills pests (the plant contains a substance that makes the insect's stomach explode). McLain was on the receiving end of the single biggest innovation to come out of the consolidated seed industry. And just as they did for the McLains, GMOs came into our universe without a history—literally.

There were no existing laws in place to govern the new creations. The technology got a real boost when President George H.W. Bush, an avid de-regulator, asked Vice President Dan Quayle to identify environmental health and safety laws for dismantlement and consider whether there should be government oversight of the new technology of genetic engineering. Those old enough to remember might recall that Quayle was chosen by the first president Bush as his running mate in 1988 primarily for his all-American good looks and supposed appeal to young voters. He came, not incidentally, from a farm state— he was a senator from Indiana—and had received hundreds of thousands of dollars in campaign contributions from large agri-chemical companies, which would soon enough dominate the seed business. He was not known for his scientific acumen, though he was known for infamously misspelling the word "potato" in a public gathering with school-children (he spelled it with an extra "e" at the end).

When the president gave Quayle the task of deciding what, if anything, to do about GMOs in the summer of 1992, the vice president reviewed a minimal amount of data; there was very little available on either the risks or benefits. Based on this cursory review, he declared that such crops were the "substantial equivalent" of food bred according to conventional breeding techniques. At their birth, genetically engineered seeds were declared by the government to be, for legal purposes, the same as traditionally bred seeds. There would be no special regulatory provisions needed.

The Food and Drug Administration initially objected to Quayle's declaration, concerned about the unintended consequences of shaking up the genome from the inside. They wanted every new GMO to be reviewed for its health implications before entering the market. But they were told by the administration to stand down.[16] As a result, there were no special requirements mandated before the mass introduction of genetically engineered crops into the US food system. The *New York Times* characterized Quayle's declaration as a "favor" to Monsanto.[17]

The coming decades would come to test the meaning of the word "equivalent."

# CHAPTER 4

# ACTS OF MAN: THE GENETIC CO-EXISTENCE CONUNDRUM

**EVERY HARVEST SEASON,** Tom Erhardson, a farmer in southern Illinois, sends off trailer-loads full of organic corn to Grain Millers, Inc, the grain processing firm located in Eden Prairie on the Minnesota River just outside of Minneapolis. If you eat grains, there's a good chance you've eaten something that's passed through Millers, which supplies powerhouses of the food industry, ranging from the ingredients for breakfast cereals to grits to multi-grain bread. They've been Erhardson's go-to buyer for the organic corn and soybeans he's been growing for the past twenty years.

"My yields are good," he told me. "And we get about double the price for my crops than if we were conventional."

Then in the spring of 2016 he sent off a fateful truckload of nine hundred bushels of corn to Millers.

Getting an official USDA organic certificate is not easy. It means more work in the field—no synthetic chemicals or chemical treatment of seeds are permitted and GMOs definitely do not qualify. Farmers have to do more monitoring for pests and weeds than in conventional fields in which chemicals do a lot of the protection work. For this special attention and extra labor, Erhardson and all organic farmers receive a premium price for their crops, ranging from 25 to 50 percent higher than conventional food.

To maintain the integrity of that distinction, Millers Grain goes to great lengths to keep sections of their processing facility separate. Located about ten miles from where the Minnesota River flows into the Mississippi, they divide the grain processing part of the facility into two sections: one is for handling crops grown with genetically engineered seeds, the other for non-engineered seeds. The two are kept in separate storage and processing facilities to prevent any mixing. As the market skyrockets for non-engineered cereals and grains, said Roger Mortenson, a grain buyer who I reached on the phone at company headquarters, they've had to expand the storage space to keep up. Keeping the two distinct, however, is getting tougher and tougher—not because of what happens at Millers but because of what happens before the crops ever get there and are sent down the Minnesota River into the Mississippi.

**THE MISSISSIPPI RIVER** has always been a central artery for American food. From these riverbanks where Huck Finn and Tom Sawyer frolicked their way into our imaginations come pouring the bounty of cereals and grains from the farms of Iowa, the Dakotas, Minnesota, and Illinois. In the river-town transit centers like Springfield, Peoria, and Beardstown, barges and small freighters bob in the water, tugging at the heavy ropes that tie them to the docks. On the land, idling trucks belch exhaust, waiting to unload tons and tons of corn and soybeans and other grains like wheat, barley, and rice. You can almost taste the bounty of grain in the thick, syrupy, pollen-heavy mists that hang over the towns and the river during harvest season.

Amidst these scenes of abundance, tucked between the granaries, the silos, the rocking docks, and the truck stops, there is something new: lab kits. Each one is little more than a beaker and a test strip, but has the power to determine the final destination of these truck-loads of plenty, as well as the fate of the farmers who grew them.

The tests take no more than five minutes to complete. A sample of each load is scooped from the back of the truck or barge, ground up and mixed with water. A strip treated with GMO antibodies is dipped into the mix. "It's like a pregnancy test," cracked Merle Kramer, who runs the Midwest Organic Farmers Cooperative in Missouri. If the strip shows a pink band, it suggests the antibodies have activated and means the load might contain genetically engineered ingredients—and a farmer's financial prospects may suddenly, and without warning, implode.

Back in Eden Prairie, the strip dipped into Erhardson's sample emerged with a pinkish hue, not good news. These spot-checks are the first round of GMO tests. They can tend toward false positives, so after the first "fail," most distributors, including Millers Grain, then use a more precise and time-consuming test that offers a final verdict. It's more expensive, around four hundred dollars per sample, compared to seventy-five dollars for the on-site strip test. And it has to be sent off-site. Following the usual procedure for suspected genetic engineering, Millers sent Erhardson's sample down the river to New Orleans.

**DAY IN AND** day out, seed samples arrive in one-pound boxes via UPS and FedEx at the headquarters of Eurofins Scientific, located on the banks of the Mississippi River just north of New Orleans. The French company is one of the biggest in the world for environmental testing.

They test for residues of pesticides in soil and food, the chemical composition of consumer products to ensure they meet US and EU specifications. And they are where the buck stops for GMOs, the final and definitive test for signs of genetic engineering inside crops before they go to market.

From the outside, genetically modified seeds look the same as their non-modified counterparts. But look inside and it's another story, according to Frank Spiegelhalter, the chief scientist at Eurofins' Gene Scan division, who invented the company's GMO test. I called Spiegelhalter at the company's New Orleans testing facility. He was genial and forthcoming, and didn't offer an opinion one way or another about the merits of genetically modified food. Speigelhalter's mandate from Eurofins when the company entered the testing market at a time of high tension between the United States and European Union over GMOs, in the early 2000s, was simply to devise a test to determine if genetic modification of a crop had occurred.

He explained what happens when a shipment of samples like Erhardson's gets to Eurofins. The kernels are ground and mixed with a chemical reagent—which has the effect of amplifying segments of the DNA inside a plant's germplasm. He or his fellow technicians then look at the gene stew under the microscope. If a genetic sequence has been altered, it will be instantly visible. "Under the microscope," he told me, "you can actually see the altered DNA sequences that are commonly used in GMOs."

As GMOs have become the dominant seeds across more than 170 million acres of US farmland, Eurofins and other similar testing facilities have been seeing those altered sequences with increasing frequency—genes where they're not supposed to be. The rapid proliferation of GMOs have occurred at the same time as another rapidly spreading counter-trend: the cultivation of organic seeds, planted as of

2016 on more than forty million acres and on the upswing. From 2011 to 2016, sales of organic food grew at least 10 percent a year, the fastest growing sector of the food market by far, according to the Organic Trade Association. It is now a $4 billion business. Dozens of new organic seed companies have launched since 2011 to feed the demand.

The two burgeoning agricultural strategies are colliding.

**BY THE TIME** Erhardson's cargo of corn arrived at the port of New Orleans—which by barge from Minnesota can take as long as one to two weeks—the definitive verdict was in from Eurofins. His "organic" corn contained over 2 percent of genetically modified elements in its germplasm, characteristics he never asked for and did not want. The news was disastrous.

The price for Erhardson's corn fell immediately. Gone were the export markets like Japan and nineteen European countries, including Germany, Denmark, France, Italy, and Spain, which have banned the import of GMOs. Gone, too, were markets closer to home, including organic lines started by big food companies like General Mills and Campbell's, or other independent lines like Kashi. Instead of selling to discerning customers willing to pay extra for no GMOs in their food, Erhardson had to settle for selling his corn at the bottom of the market as animal feed, where the prices are barely half of what he would command for a non-genetically engineered crop. He lost more than nine thousand dollars on that single trailer-load of corn and, when I spoke with him in 2017, he is afraid it could happen again.

Nor is "Erhardson" the farmer's real name. He requested that I not use his name because of fears that *he* could be sued by Monsanto or another seed company for having inadvertent GMO content in his crop. Nor is he by any means alone; Erhardson's story is being repeated daily across the agricultural lands where GMOs and organic crops are being grown in the vicinity of one another. More and more farmers are alarmed to find that the genetic material they have been purposely avoiding is turning up in their crops.

Monsanto has stated publicly that it allows for the possibility of accidental contamination and will not prosecute any farmers with as much as one percent contamination. But two percent, or even more? Erhardson, like many other farmers in the area, does not want to be a test case.

Merle Kramer, director of the Midwest Organic Farmers Cooperative, has no such fears, because he can talk in the aggregate about the experiences of the farmer-members of his co-op in Iowa, Wisconsin, Missouri, and Illinois. Kramer estimates that at least a third of the co-op members have experienced contamination of their organic crops over the last three years. The contamination translates into at least half a million dollars every year, he told me—a significant loss for farmers already on a narrow margin. And Kramer's is just one of many farmer coops throughout the Midwest that have reported contamination.

Corn is highly promiscuous, whether it's genetically engineered or not. Left to its own devices, when ripe, the plant's fluttering tassels send pollen to wherever a receptive female awaits, whether it be in the plant next door, the farm next door, or the organic field next door. Genetically engineered DNA can also come in through kernels left in shared farm machinery or trucks—which is why most non-GMO granaries require a "truck-wash" certificate (for fifty bucks) affirming that a truck carrying non-GMO crops has been cleansed of any previous crop shipments.

Erhardson told me what he thinks happened on his farm. Normally he plants his summer corn seeds two to three weeks later than his GMO neighbors, so their corn doesn't go into pollination at the same time. But that year, it was a little cooler than previous years. It took longer for his neighbor's corn to ripen, which meant the tassels on the corn he planted became receptive to pollination at just the time when his neighbor's corn was ripening. All that pollen, he surmised, came blowing in the wind and mixed with his own carefully tended organic corn plants.

"His pollination window was still open at the same time mine was just opening up," he said.

And so, the corn in that field in southern Illinois suddenly bore the characteristics of the genetically altered field next door. And the ancient primordial process by which plants reproduce themselves by sending flying pollen off into the wind to find a willing partner can derail a farmers' financial prospects, undermine their sense of dominion over their own land, and create suspicion about neighbors' farming practices. "Blowin' in the wind," an image celebrated in poetry and song, now comes with sinister implications.

The USDA's Economic Research Service estimates that in the 2014 crop cycle, organic farmers lost at least $6.1 million in sales when the price for their crop plunged due to genetic contamination, averaging about $71,000 per contaminated farm.[1] That is a very low estimate of the financial damages from contamination, since there is no official mechanism for registering such events. It doesn't account for the expenses incurred by farmers who try to insulate themselves from contamination—like the land taken out of cultivation to create buffer zones between GMO and non-GMO fields, and the staggering of planting and harvest schedules to avoid cross-pollination. The latter strategy, according to Kramer, often leaves non-GMO farmers with less wiggle room to time their harvests with highs in the market. The Government Accountability Office criticized the USDA for inadequately measuring or monitoring the multi-level consequences of contamination and its inability to capture meaningful data on the scale of the problem. The agency, it concluded, "lacks statistically valid data needed to understand the full scope of the potential economic impacts from unintended GE presence."[2]

Out in the field, the full scope of contamination is not ambiguous. More than 85 percent of farmers surveyed by the Organic Farmers Agency for Relationship Marketing, which promotes organic agriculture, and Food & Water Watch, an agriculture policy watch-dog, said they were "very concerned" or "concerned" about GMO contamination affecting their farms. Organic farmers spent an average of $4,500 apiece just to try to protect their farms from GMO incursions. That may not sound like a lot, but add it to the lost markets and evaporating premiums on farms that are usually smaller than conventional farms and operating on already narrow margins and the consequences rapidly add up.[3]

"The whole thing is nerve wracking," John Bobbe, executive director of O'FARM, told me. "Contamination puts our farmers in the middle of the shooting gallery."

**WHO PAYS FOR** the losses when organic farmers find unwanted genes in their crops? GMO patent holders fought every effort to take financial responsibility for the losses due to their product. The Organic Seed Alliance and O'FARM have repeatedly requested that the USDA devise ways to ensure farmers are compensated for their lost markets and for the costs of protecting themselves from "the seeds next door." "Financial responsibility," asserts Kristina Hubbard, policy analyst and organizer for the Organic Seed Alliance, "belongs on the shoulders of the manufacturing companies who profit off those (GMO) varieties." Farmers began showing up at USDA meetings to demand action. Pressure mounted.

Finally, Secretary of Agriculture Tom Vilsack took action. He appointed a committee and gave it a grand title, the *Advisory Committee on Biotechnology and 21st Century Agriculture*, which came to be known in ag circles as AC-21. Their mandate was to recommend how to, "strengthen coexistence among different agricultural production methods." In other words, figure out how GMO and non-GMO farms could coexist within proximity of one another.

The committee spent four years on the conundrum. One idea would have imposed a small levy on GMO farmers to create a funding pool that could be used to compensate victims of contamination. Others suggested pooling insurance funds that would be financed primarily by the companies that produce, and profit from, GMOs. All were strongly opposed by the bio-tech industry and went nowhere. In one revealing example, the committee invited a USDA official to discuss whether crop insurance policies could be tailored to protect farmers against financial losses due to contamination. He vehemently rejected the idea. His reasoning, delivered on power point, according to a transcript of the proceedings, was revealing. The agency could not insure against GMO contamination because insurance normally protects against so-called "acts of God"—such as drought, heat, and floods. Over those factors, he said, humans have no control (though all of these phenomena are increasing due to climate change). By contrast, GMO contamination, he told the committee, is clearly a "manmade" problem. He could not envision writing an insurance policy for practices that are, given the right set of circumstances, made "inevitable" by the acts of humankind.

AC-21 issued its final recommendations in December 2016.[4] There would be no compensation. Farmers, the blue-ribbon committee suggested, should talk to each other more, and coordinate planting schedules and expand buffer zones. All such efforts would be at the non-GMO farmers expense and initiation. Farmers I spoke with could barely muster a discouraged shake of the head at how completely the council's ambitions to tackle rampant contamination had collapsed into "talking with your neighbor."

"All of the costs of prevention, all of the costs of contamination," Patty Lovera, assistant director of Food & Water Watch, recollected, "all of that falls on the non (GMO) users' side of the fence."

**AFTER A TWENTY-YEAR** experiment in our fields and in our kitchens, GMOs are deeply implanted in the American food system.

When they were first introduced, everything about genetically engineered seeds was new. Genes from trout were being spliced into strawberries, genes from bacteria spliced into potatoes, genes from hogs spliced into corn. There was no "prior art," a term the Patent Office uses to weigh the originality of a patent applicant's invention. They were, indeed, original. Almost every GMO patent application— and there was a deluge of them starting in the early 1990s—was approved. Which, according to Andrew Kimbrell, executive director of the Center for Food Safety, creates two competing strands in the GMO story, between what the companies tell the Patent Office they are and what the government tells the public they are.

Kimbrell has read through dozens of GMO patent applications, and in some instances challenged them in court. "On the one hand, the companies want to claim that nothing like this has ever been done before, that for patent purposes it is original and different," he commented. "On the other hand, they say to the regulatory authorities it's the 'substantial equivalent' of other crops." So patents are granted on the basis of their difference from other seeds while their regulatory status is based on the principle that they're no different from other seeds.

What we know for sure is that GMOs look different under a microscope. Their genes are mobile. They carry financial implications because of consumer concerns.

The question is: Does it matter? Does engineering the genetics of seeds with genes from unrelated species make a difference on the farm or for those who consume them? Are they dangerous to our health? What's going on over there, on the other side of that fence?

# CHAPTER 5

# GENETIC ROULETTE: ENGINEERING THE SEED

**THE MONIKER OF** "substantial equivalence" has been expensive to sustain.

The money has been rolling in.

Walk through a supermarket and GMOs are everywhere: they're in canned vegetables, canned fruit, most processed cereals, processed food mixes, snacks, candy, corn muffins, corn chips, corn tortillas, corn transformed into fructose and syrup and oils and adhesives; they're in most brands of canola oil; they're in canned beets; in a lot of Hawaii-grown papaya; in multiple derivations of soybeans; and, up and coming, on increasing acres of test plots approved for genetically engineered squash and potato,[1] apples, peaches, and pears.[2] Chances are, if it doesn't have a label saying it has not been genetically modified—i.e., "GMO-free" (which means it's gone through testing like Erhardson's crop)—then there's a good chance that it has been.

But we can only surmise. We still don't know how much genetic engineering has intruded into the food we eat because of the extraordinary efforts made to prevent the public from knowing. The biotech and grocery industries—through their trade groups the Biotechnology Industry Organization and the Grocery Manufacturers Association, respectively—have spent at least $143 million to oppose state labeling initiatives and ballot measures in Oregon, Washington, Colorado, and California in just the two years between 2013 and 2015, according to the Environmental Working Group.[3] Between 2009 and 2013, the biotech trade group BIO poured another $314 million into supporting research, lobbying, public relations, and candidates to promote the expanded use of GMOs and oppose measures to restrict them.[4]

The industry supported a far weaker version of labeling, which passed Congress and was signed by President Obama in 2016. The Safe and Accurate Food Labeling Act (nicknamed by opponents the Denying Americans the Right to Know Act, or DARK), eliminates the ability of states to pass labeling laws and instead requires that GMO status be included in a phone app; wave your phone like a wand over a produce or fruit stand and it's supposed to identify whether or not the goods have been genetically altered. And if that doesn't work, or if you don't have access to Wi-Fi, there's supposed to be an 800 number created to call for more info—all while whirring through a market looking for food. One of the measure's leading advocates was then-Republican Congressman Mike Pompeo of Kansas, who received more than $100,000 from agri-business and food companies for his campaigns, and was selected by President Trump to direct the Central Intelligence Agency, and, later, Secretary of State.[5] Altogether, a total of almost half a billion dollars has been devoted to promoting a product to the public which is officially the equivalent of all other foods, while weakening the public's ability to be informed when we consume them.

In the latter half of 2017, however, a change occurred: half-a-dozen food firms withdrew their membership in the GMA, moves widely considered to be related to the trade group's ardent and vocal opposition to labeling. The companies include giants of the food business—Campbell's, Unilever, Mars, Nestle, and the commodity giant Cargill. All are trying, to varying degrees, to re-position themselves with a public perceived to be more attuned to the environmental backstory to their food sources than ever before.

Long before labeling became the cutting edge of disputes over GMOs, in the mid-1990s it was government money that started rolling in: hundreds of millions of public dollars poured into universities and scientific institutes to study new ways to do genetic engineering for crops, planting and water systems to conform to their needs, and ways to make them amenable to long-distance transport and long-term storage. A fraction of those funds went to fund research into potential GMO risks. One example, from 2016: in that year the USDA's Biotechnology Risk Assessment Program was granted a total of $4.4 million—its highest total yet—to study the risks to human health or the environment from GMOs. In that same year the USDA's National Institute for Food and Agriculture awarded grants amounting to more than $15 million for research into how to more efficiently and productively cultivate genetically engineered seeds. That fourfold discrepancy has been characteristic of the priorities of Democratic and Republican administration's alike; in fact the gap has usually been far wider. In 2017, President Trump signed an Executive Order that may be the most aggressive effort yet to promote GMOs in the United States and abroad—in the name of "rural economic development," and to convince skeptical foreign governments, of which there are many, to retreat from regulating their use.

Money has also been channeled to sympathetic scientists, contributing to potential conflicts of interest. At least 40 percent of the scientists involved in more than six hundred research studies on GMOs published between 2005 and 2015 were found by the peer-reviewed journal *PLOS-One* to have financial ties to bio-tech companies, including research contracts and patent deals. Those financial ties increased the chances by 50 percent that the outcome of "research" into the effectiveness of genetically engineered Bt corn would be positive.[6] Even the Academy of Sciences's Natural Research Council was found to have numerous un-reported conflicts of interest among the committee members who authored reports on GMOs.[7]

Funding also has gone to undermining the organic competition. When newspaper articles appeared in 2016 citing a scientific study suggesting that the organic industry was overstating the nutritional and other benefits of chemical-free food, most of the sourcing went back to a group called Academic Reviews, ostensibly formed to disseminate expertise on "agriculture and food science." After many headline grabbing stories—*"Organics exposed!"* etc.—U.S. Right to Know, which specializes in doing deep dive Freedom of Information Act requests, revealed that a significant part of the funding for Academic Reviews came from Monsanto. "They have an obvious interest in undermining organic, which prohibits use of their treated seeds or chemicals," commented Stacy Malkan, who unearthed the documents and is co-director of U.S. Right to Know.[8]

So, GMOs and the companies that own them have been favored with taxpayer funding and with a significant public relations effort on their behalf. Back to the question: Do the distinctions between GMOs and conventional crops make a difference?

**ON THE MATTER** of whether or not genetically engineered foods are safe to eat, there is still little conclusive evidence in either direction. But concerns are rising. Sheldon Krimsky, a professor of Public Health and Family Medicine at Tufts University, reviewed twenty years of studies in which animals were given GMO feed, and concluded that they raised questions yet to be resolved about the potential health impacts of consuming GMOs.[9] These focus primarily around lingering uncertainties over how GMOs are metabolized by the micro-biota of the gut; and on the potential transfer of allergenicity from one type of plant to another. For example, if you're allergic to peanuts and a peanut gene is transferred to a soybean, would that make you allergic to the soybean? For these reasons, significant parts of the health profession have come out in favor of labeling on the principle of informing the public, including the American Nurses Association, the American Public Health Association, and the state medical societies of Illinois, Indiana, and California.

What's most telling about Krimsky's findings is how indefinite they are. This is the scientific method at work in real-time: the study is useful not necessarily in providing a definitive answer—*are they safe?*—but in identifying how little is known, and what areas in that realm of the unknown could be ripe for further research. In a further sign of scientific unease, three hundred of the world's top environmental health scientists and doctors signed a letter to the journal *Environmental Sciences Europe* in 2016 asserting that it is the uncertainties about GMOs and the inadequacy of existing data for making any kind of safety determination that troubles them. Their letter was headlined, "No Scientific Consensus on GMO Safety," and was signed by scientists from leading institutions in the United States, Canada, the United Kingdom, Germany, France, Japan, New Zealand, and elsewhere. It's critical to note that the scientists did *not* state that GMOs are unsafe. Rather, they argued that, " . . . the scarcity and contradictory nature of the scientific evidence published to date prevents conclusive claims of safety, or of lack of safety, of GMOs."[10] In other words, there is no clear evidence that eating genetically engineered food is harmful to your health, nor is there clear evidence that it is safe.

When it comes to the impact of cultivating GMOs on the agricultural economy, farm ecology, and health of farmers and farmworkers, however, the evidence is accumulating. It suggests that GMOs are directly and indirectly making our food system more precarious and undermining its resilience.

Here's where the evidence leads, in five points in five sentences. In short: the spread of a uniform set of seeds across huge parts of the United States is encouraging monocrop farming and leading to the narrowing of seed diversity that scientists warn could lead to significant crop losses. Insects and weeds are developing resistance to ever-higher volumes of chemical poisons tied closely to the cultivation of genetically engineered crops. Partly as a result, the yield bonus that was promised when GMOs were introduced has not materialized. Serious concerns are being raised about the public health consequences of the chemicals used to sustain GMOs in the field— including the world's most popular herbicide, glyphosate. And the technical requirements for creating a GMO seed are so capital intensive that the effect is to concentrate evermore power in the few companies which can afford to produce them.

Let's unpack each of those areas of concern.

## NARROWING OF SEED DIVERSITY

Across a landscape that once hosted hundreds of different crops and varieties, genetically engineered seeds are drawn from a narrow genetic pool. Indeed, the vast expansion of GMOs corresponds with a plunge in diversity of America's seed stock—most prominent with corn and soybeans, and increasingly the case with canola, sugar beets, and alfalfa (food for dairy cattle). More than 90 percent of corn and 75 percent of all soybeans in the United States—as well as smaller portions of canola, sugar beets, papaya, alfalfa, potatoes, and cotton—are planted with genetically engineered seeds.

The Kansas and North Dakota university study[11] that demonstrated declining diversity on American farms starting in the early nineties correlates with the emergence of genetically engineered seeds as they came to dominate agriculture in the latter years of the 1990s. Such narrowing of the genetic base, they concluded, "could have far-reaching consequences for provision of ecosystem services associated with food system sustainability."

Traditional breeding, like that which Nikolay Vavilov and Luther Burbank and their many successors pioneered, involves enhancing characteristics generation upon generation until breeders get what they're looking for—hardiness, resistance, enhanced taste, etc. Creators of genetically modified varieties short-circuit this time-consuming yet critical process. Instead, they seek out one trait in a taxonomically un-related organism, identify the responsible gene, and transplant it.

Consider where many of the GMO seeds are devised and tested—in sub-tropical Hawaii, a world away geographically and ecologically from the American Midwest where most of them are cultivated into crops. Islands in the middle of the Pacific Ocean are the key breeding and testing grounds for GMOs. About an hour's drive north from Honolulu, on plots practically around the corner from the US navy base in Kunia where Edward Snowden worked before revealing how the United States was secretly eavesdropping on the world's conversations, genetically engineered seeds destined for American farm fields are germinated and tested. Monsanto, Syngenta, and DuPont Pioneer established their forward operating bases on Oahu's North Shore, as well as on the neighboring island of Kauai. Seeds that germinate amidst the reeds and palms of Hawaii are shipped off from here as seedlings for Midwestern farmers. Their presence has generated fierce opposition from residents, who have objected to the agri-chemicals being used on the germinating seeds, the details of which the companies have refused to release. Citizens succeeded in passing referendums and regulations banning the cultivation or open-air testing of GMOs on the islands of Maui, Kauai, and the big island of Hawaii. Those measures have been fiercely opposed by the industry, and overturned in court on jurisdictional grounds holding that local authorities could not pre-empt state law, which permits their use.

The companies love Hawaii because it boasts as many as four growing seasons, Monica Lynn Ivey, a spokesperson for Monsanto based in Kauai, told me. They also take advantage of the weak oversight of agri-chemicals in the state, which are a critical part of the GMO package. The lab-engineered seeds—many of them designed in those Hawaii labs and test plots—are insulated from the normal call and response mechanisms of evolution in the field. The fact that conditions on the humid and tropical Pacific island bear no relation to conditions in the dry Midwestern plains where most of the seeds are grown out into crops enhances their reliance on genetic manipulation and powerful chemicals to survive.

Most genetically engineered seeds express just two traits. One is obtained from a soil-borne bacteria, called Bacillus thuringiensis, or Bt, which contains a poison that is toxic when consumed by certain insect pests; and another is obtained from a virus that makes crops resistant to glyphosate. Seeds which are engineered to be resistant to chemicals, as is the case with Roundup Ready, or create their own, as with the 'Bt' trait, will not experience the evolutionary cycles that can lead to pest or weed resistance within the plant. Instead, they're delivered with an added gene from another species. One commercial breeder told me a revealing bit of jargon from the plant universe: seeds that require a high quantity of nitrogen-based fertilizers and synthetic chemicals to survive are known, informally, as "the babies," because they're not tough enough to produce food without the inputs.

"They're selecting against competitiveness," said Doug Gurian-Sherman, senior scientist at the Center for Food Safety. Gurian-Sherman has a PhD in plant pathology and worked for several years in the early nineties in the Biotechnology Group at the US Patent and Trademark Office. "Engineering a piecemeal gene here or there, in a plant that has 30,000 or 40,000 genes, is a technological steam roller substituting for the benefits of diversity."

Fear that the narrowing of seed stocks might make America's food system vulnerable to collapse like a house of genetically identical cards finally caught the attention of Secretary of Agriculture Tom Vilsack. In 2014, Vilsack requested a study on the genetic composition of eight crops that had a significant quantity of genetically engineered seeds—corn, soybeans, canola, alfalfa, beets, squash, papaya, and cotton. A committee of experts—scientists, farmers, and seed company executives—called the Genetic Resources Advisory Council (GRAC) was convened. They were asked for what amounted to an update of that genetic map of our seeds compiled by the National Research Council in the wake of the 1970s-era corn blight. At the very least, Vilsack hoped they could identify crops in which genetic vulnerability was of highest concern. By 2015, however, the GRAC ran into a wall—namely, American patent laws.

Bill Tracy, chairman of the GRAC's Corn Germplasm Committee, was hoping to get access to some of that corn germplasm so the committee could study it, and was rebuffed by the companies that controlled the patents. Even after offering strict privacy protocols, he told me, the committee was refused access to many of the seeds on the grounds that their genetic information lay behind trade secrets. Other crops faced similar obstacles. The inquiry was stopped in its tracks. Tracy has a disturbing theory as to why.

"Why are the companies so concerned about us studying their germplasm?" Tracy asked, still exasperated at how he and others had been stymied in their research effort. "I think it's because they're more uniform than we suspect."

Tracy suggested I have a look at a study published in the journal *Crop Science* for some clues into the companies' secretiveness. An agronomist at the University of Illinois, Mark Mikel, found that most of the corn seeds grown in America are based on a narrow spectrum of genetic material, narrower even than that which triggered the 1970 corn blight. In fact, Mikel discovered that the most popular corn varieties are descended from the same parent-lines: there was a "high genetic contribution" of Pioneer germplasm in Monsanto's varieties, he concluded, and numerous overlaps between the traits expressed in the Pioneer and Monsanto seeds. The two companies, he concluded, "are in practice primarily breeding new germ-plasm through recycling closely related inbred lines within their respective germ-plasm pools." Translation: the country's two largest and ostensibly competitive seed companies are drawing from the same limited gene pool, resulting in seeds with highly similar genetic personalities.[12] Mikel's findings comport with revelations from the Farmer Business Network—which cleverly detoured company controls by simply asking for farmer seed samples—that many differently labeled varieties are actually the same seed.

"The companies have a great deal of secrecy," commented Major Goodman, the veteran plant scientist at North Carolina State University. "Part of that secrecy is because of real trade secrets, and part of that is embarrassment for what they have. They are all very much the same, no matter what the company. Look through where the pedigrees lead and you see why certain companies are not interested in being open to a genetic assessment."

The inability of farmers and the government to access the composition of genetically altered genes has left us without a clear picture of our genetic vulnerabilities. We are flying blind, or more precisely, planting blind. All indications are that we're relying, for our major crops, on a thin thread of genes that's getting thinner just as we head into conditions that require a more expansive genetic foundation.

## INSECT AND WEED RESISTANCE

Genetic engineering may halt the evolutionary process, but among the creatures that accumulate around and feed off crops, evolution does not stop. In fact, like any ecological process, it continues: the corn borer is developing resistance to the toxins inside each genetically engineered Bt plant designed to kill it,[13] and resistance to Roundup has become widespread. The major global association of weed scientists, representing experts in eighty countries, identifies twenty-four species of weeds that are now wholly immune to the effects of Roundup's active ingredient, glyphosate.[14] New generations of super-weeds are proliferating across some sixty million acres of soybean fields that no longer respond to the toxins in Roundup, reports the Union of Concerned Scientists.[15]

The result is a chemical arms race in the making—against weeds. Ever-greater individual doses of Roundup are being used, and even more potent chemicals being devised. Monsanto has begun marketing a new generation of seeds stacked with traits to resist as many as three herbicides at once. That includes Dicamba, a potent herbicide that is also prone to drift: several lawsuits have been filed by farmers in Arkansas, Missouri, and numerous other states alleging that their crops, which do not contain the engineered gene making soybeans immune to the potent herbicide, were destroyed when it was sprayed on fields by neighboring farmers. Arkansas and Missouri have actually banned the use of the herbicide anywhere in the state. Dow has developed a new package of seeds designed to resist applications of 2,4-D, which was an ingredient in the notorious Agent Orange concoction used by the US military during the Vietnam War to defoliate Vietnamese forests. The herbicide has a similar effect on weeds in the field that its more potent cousin had on the rainforests of Vietnam, denuding all in its path, as well as being deemed a "possible" carcinogen by the World Health Organization. It's also highly persistent in the atmosphere, triggering farmer protests by a Save Our Crops coalition across the Midwest.

Meanwhile, research suggests that such blanket chemical applications would likely not be necessary if locally evolved seeds—known as "historical cultivars"—had not been replaced by ones bred to be dependent on the herbicide to begin with. Such "historical cultivars," perform at least as well resisting weeds as do genetically engineered varieties, according to a research study published in the *Journal of Chemical Ecology*.[16] "Historical cultivars have better weed suppression ability than modern cultivars," the scientists conclude, and suggest that mutations which occur naturally in a crop population evolving over generations in a specific eco-system are better equipped to resist familiar and unfamiliar threats than engineered varieties designed to express just one trait. That one trait, of course, has nothing to do with strengthening the plant to resist weeds, but to resist the toxic impacts of a chemical weed killer. Meanwhile, it's the weeds that are learning to resist the weed killer.

## LACK OF GMO YIELD BONUS

From the beginning, genetically engineered seeds were presented as something of a magic bullet to increase food production for a rising global population. Genetic engineering, according to the industry's trade association, BIO, will "increase crop yield and farmer income."

The National Academy of Sciences set out to document the overall impacts of genetically engineered crops. One matter they addressed was whether the new technology resulted in increased yields beyond those that would have happened anyway. Yields per acre were already increasing before the onset of GMOs, so a key question was whether the yields of genetically engineered crops were increasing at a quicker rate than the yield increases of conventional crops. Did the use of GMOs make a difference in the quantity of food grown? The Academy's answer was contained in a 380-page report released in May 2016: there was no discernible difference between the two. GMOs, they said, had neither a negative nor positive impact on yields when compared to the performance of conventional (i.e. non-GMO) crops.

There was no evidence, they reported, "that the average historical rate of increase in U.S. yields of cotton, maize and soybean has changed."[17]

The Academy also considered the association of GMOs with agrichemicals. They noted two contrasting trends: herbicide use has dramatically increased on genetically engineered soybeans, but on corn fields which contain the engineered Bt bug poison, pesticide use has significantly decreased (a trend that is slowing as pest resistance grows).

Several months after the Academy weighed in, the *New York Times* followed up with a detailed investigation into the yield question, and compared the relative productivity of farms in the United States and Canada, where the use of genetically engineered seeds is widespread, and farms in France and Germany, where they are, for the most part, forbidden. The newspaper also found no difference in the rate by which yields increased on farms growing genetically engineered corn in the United States and non-GE corn in France; the same was true for rapeseed, used for canola oil.[18] Similarly, a study from the University of Canterbury, New Zealand found that corn and canola yields were actually rising much quicker on non-GMO farms in western European countries than on farms using GMOs in the United States and Canada. They surmised that, "Europe has learned to grow more food per hectare and use fewer chemicals in the process. The American choices in biotechnology are causing it to fall behind Europe in productivity and sustainability."[19] The study helped the government of New Zealand determine whether the country would follow the United States or the far warier European approach to GMOs. They opted for the latter; New Zealand maintains one of the world's most rigorous regimes governing the importation or cultivation of GMOs.[20]

## HEALTH CONSEQUENCES?

While yields have shown few increases that can be clearly attributed to genetically engineered traits, what has shown skyrocketing increases are applications of the herbicide glyphosate, aka Roundup. As of 2016, the equivalent of about a pound of glyphosate is applied to every cultivated acre in America. The use of Roundup, and most recently its generic equivalent, has skyrocketed along with the emergence of GMOs—from about twenty-eight million pounds yearly over the four-year period between 1992 to 1996, to an average of about seventy-nine million pounds yearly between 1996-2000, the years when Roundup Ready soy, corn, and other crops spread across the land; and it's been on a steadily upward incline since.[21]

The chemical that hitchhiked its way on GMO seeds into the center of the American food system was developed by Monsanto in 1974 to kill weeds by interfering with the production of an enzyme that is key to a seed's ability to grow into a plant. Genetic engineering creates a seed with a trait of immunity to Roundup's toxic effects, so that farmers can kill unwanted plants—weeds—without killing their crop. That immunity to Roundup is now featured in an expanding number of Monsanto's genetically engineered seeds, including corn, canola, sorghum, winter wheat, alfalfa, and cotton.

Glyphosate is the not-so-secret ingredient behind Monsanto's rise from a small, regional chemical company—it was founded in St. Louis in 1905 and produced nerve gas for the US Army during World War I—into a key player in the chemical-seed nexus. The substance is key to the company's financial performance.[22] Between 1997, when Roundup Ready crops were first planted on a large scale, and 2008, when the company's patent ran out, glyphosate alone generated as much as a billion dollars in profits annually.[23] In 2016, sales of glyphosate still represented 26 percent of Monsanto's total revenues of $13.5 billion, according to the company's annual report; and that doesn't include the $8 billion in sales of corn and soybean "seeds and traits" (which is how they're described in the company's annual report), many of which are bred to require glyphosate and other chemical applications. The chemical is now applied to about two-thirds of all genetically engineered crops, amounting to about 825,000 tons annually on eighty million of the world's cultivated acres.

It's fair to say that glyphosate has been central to Monsanto's rise and to the reshaping of the seed business into a multinational oligopoly of chemical companies. The chemical has been so central to Monsanto's rise that it got the company into trouble with the Securities and Exchange Commission (SEC). In 2015, the SEC accused the company of neglecting to report to shareholders millions of dollars in payments it made to seed and ag-chemical outlets to favor Roundup over less expensive generic brands made by the company's competitors. According to the SEC complaint, the unreported payouts to US and Canadian retailers from 2009 to 2011 amounted to $194 million; $24 million to French retailers; and another $10 million to German retailers—for a total of $228 million.[24]

"They were paying off dealers to keep Roundup in their inventories," commented Philip Howard, an associate professor at Michigan State University in East Lansing, and author of a seminal study on concentration in agriculture.[25] According to the SEC, the company neglected to account for the costs of the rebates to shareholders. Against all principles of the free market, Monsanto was alleged to have paid seed and chemical retailers to ensure that its weed-killer would continue to dominate in the world's GMO soybean fields. The company neither admitted nor denied the charges, and in February 2016 paid an $80 million fine—$22 million of which went as a reward to a former Monsanto executive who alerted the Commission to the accounting abuses.

Glyphosate has been Monsanto's juggernaut; it's the reason that the once-small firm from St. Louis is now considered one of the big four of the chemical-seed conglomerates. The fate of glyphosate, now appearing in generic as well as Monsanto's trademarked form, is tied directly to the fate of the world's GMOs. No glyphosate, no Roundup Ready seeds.

So it was with great displeasure, to put it mildly, that Monsanto confronted the news which landed like a grenade in March 2015: the World Health Organization's International Agency for Research on Cancer (IARC), the world's preeminent institution of cancer doctors, environmental health specialists, and medical researchers, determined that glyphosate is a "probable carcinogen," which is the cautious institution's highest level of concern. The head of the IARC assessment explained publicly that the agency based its decision on three strands of medical evidence and epidemiological surveys: there was "clear evidence" linking glyphosate exposure to cancer in laboratory animals; "strong evidence" that at a cellular level glyphosate is genotoxic, meaning it can damage human genes; and more limited strands of evidence showing cancer in humans in "real-world exposures"— which mostly means workers who apply the substance in the field. The agency identified a "probable" correlation between glyphosate exposure and Non-Hodgkin's Lymphoma,[26] a cancer that begins in the lymph system and can spread to other parts of the body, including the liver, bones, and brain. This was the agency's primary area of concern; there were few indications of links to other cancers.

IARC's action set off a frenzied chain reaction that quickly elevated glyphosate and its partner seeds into the center of a global struggle over who gets to determine what is a risk to our public health.

The dominoes began falling in the state of California, where a law requires that the public be informed when chemicals are present that cause cancer, birth defects, or harm to the reproductive system. Known as "Prop 65," the law is named after a proposition which voters approved in 1988. Across the state, signs in offices, businesses, and shopping malls declare the presence of such dangerous chemicals. The IARC is one of the few public health institutions designated as a source for the chemicals that warrant being listed. (The others are the Environmental Protection Agency, the Federal Drug Administration, the National Institute of Occupational Safety and Health, and the National Toxicology Program at the Department of Health and Human Services.[27]) Three months after the IARC finding, California issued a notice of intent to add glyphosate to the list of "Prop 65" chemicals that present a danger to the public. This action kicked off a historic battle as Monsanto sought to defend the chemical and undermine the authority of the world's premiere public health organization.

Before the state could print up new signs, Monsanto sued California to block it from listing the chemical as a danger to human health. The company attacked IARC, claiming in its brief that the state improperly delegated legal authority to "an unelected and non-transparent foreign body that is not under the oversight or control of any federal or state government entity."[28] The IARC is part of the World Health Organization and is based in Lyon, France. For more than four decades it has been a primary source of independent science on the impact of environmental toxins on human health. In March 2017, a federal court judge rejected Monsanto's claim. Monsanto appealed, but the state forged ahead and added glyphosate to its public listing of carcinogens in June 2017.

Disputes over IARC's finding then got hotter. An investigation by Reuters asserted that an environmental health scientist claimed his un-published work, which found glyphosate to be safe in limited doses, was not included in the IARC analysis.[29] Then U.S. Right to Know issued a rebuttal of the Reuters report, calling into question the journalist's credibility, and identifying numerous sources excluded from her reporting.[30] The premiere French newspaper, *Le Monde*, jumped into the fray with a two-part investigation into the industry efforts to influence Reuters and to undermine the IARC and the credibility of scientists associated with the glyphosate finding.[31]

More than forty lawsuits have been filed by farmworkers and gardeners alleging that Monsanto failed to warn them of glyphosate's allegedly cancer-causing effects. They represent at least 140 individuals who worked on farms or fields where glyphosate was routinely sprayed over many years, and have been afflicted with Non-Hodgkin's Lymphoma. By early 2017, discovery in the cases unearthed hundreds of pages of documents, many of them revealed by US Right to Know reporter Carey Gilliam, whose book *Whitewash* probed into the hidden history of Monsanto and glyphosate.[32] That included email correspondence between a top EPA official and a company official shortly before the EPA blocked an investigation into glyphosate's toxicity and declared it safe for humans.[33] In other emails released via the legal cases, it appears the company did not disclose its relationships with some of the scientists involved in studies claiming the chemical is safe.[34]

So the battle over glyphosate is in many ways turning into a proxy for the battle we never had when GMOs were first launched onto the fields of America and the world. In Congress, Monsanto has been lobbying to rescind the US financial contribution to IARC, which has ranged between one and ten million dollars annually. In Europe, more than one hundred top scientists from around the world, including the retired director of the carcinogenicity panel at the US National Institute of Health National Toxicology Program, sent a letter of protest after the European Union authorized glyphosate for use in 2016. Their eight-page letter asserted that the determination of glyphosate safety was based on a selective look at the evidence and conflicts of interest among chemical industry-linked executives on an advisory linked to the decision. A coalition of international environmental health NGOs, including Greenpeace and Friends of the Earth are campaigning to reverse decisions by the European Union and US EPA to permit glyphosate's continued use.

**WITH EVERY PASSING** day and with nearly every accumulating study, GMOs are looking less and less like the "significant equivalent" to conventional crops that they were declared to be more than two decades ago. It can cost upwards of $130 million to develop a genetically engineered seed, which is one reason why their expansion into the food system has been accompanied by a tightening of control over seeds. The expense and complexity of producing genetically engineered organisms, and the aggressive patent prerogatives needed to defend them, led to a concentration of power within a secretive group of companies over the most basic element of our existence—seeds, and the food from which they grow.

All these factors have pushed us toward a food system that is less diverse, more insecure, and delivering increasingly consequential environmental costs. "We're sitting ducks," says Major Goodman, the plant scientist at North Carolina State University. "The level of uniform germplasm throughout the world, we're just waiting for some bacteria or fungus to find a spot of vulnerability. It's not a matter of if, but when. Eventually it will catch up with us."

But maybe not.

Mention the word "seed" and people most often jump to the word "Monsanto." I heard this many times when describing this book—*"oh seeds, Monsanto, right?"* Somehow a single company succeeded in becoming the go-to association for what is the quintessential life force, often no bigger than a fingernail, that lies at the root of our survival. But a vast network of people around the world, speaking multiple languages, living in highly different places and situations, are connected by their determination to reclaim the seed from the hold of that or any other company.

# CHAPTER 6

# SEED REBELS

**FOR MORE THAN** thirty years, the Iraqi city of Abu Ghraib was renowned as the home of something with far more planetary significance than the location of a prison where US troops abused their prisoners. The city was host to one of the world's most valuable seed banks, a storehouse for some of the oldest seeds on earth. Within its simple concrete walls were seeds representing more than 1,500 varieties of food crops, including wheat, barley, fava beans, cowpeas, sorghum, chickpeas, and dryland fruit trees, dietary staples for hundreds of millions of people. Just as we humans contain the genetic information of those who have gone before us, many of the seeds at Abu Ghraib dated back as long as ten thousand years, to the first domestication of agriculture in the Fertile Crescent.

The Fertile Crescent is a broad plain of historically fertile land, arising out of the delta of the Tigris and Euphrates rivers, and stretching through what are now parts of Iraq, Syria, Lebanon, Israel, Iran, Jordan, and Palestine, and into North Africa. It's one of the most valuable regions on the planet for seeds because of a unique convergence of features: it's a hotbed of biological diversity and a center-point of origin for many of the world's most important food crops, where many of the foods eaten in North America, Europe, and elsewhere start their long path to our plates. Scattered throughout is an abundant spectrum of wild, non-domesticated relatives of food crops, so it is still possible to find ancient, tough genetic codes that can be bred into our more modern varieties. And, of utmost significance moving forward, it is a region that many of our planet's food growing lands are starting to resemble.

Among the many tragic outcomes of the ill-fated US invasion of Iraq in 2003 was the looting and destruction of the Abu Ghraib seed bank. First looters attacked, apparently interested not in the seeds but in the glass jars that held them. Iraqi scientists and technicians made a heroic effort to gather up the seeds they could; many of them had to be scraped off the floor where they were spilled from pilfered glassware. Then they spirited the seeds out of the country before the facility was destroyed by either an American bomb hitting the unintended target (a frequent occurrence) or an Iraqi explosive device doing the same. They rushed the seeds off to what they hoped was a new secure location across the border—to the Syrian town of Tel Hadya, about twenty miles from the city of Aleppo.

Tal Hadya was home to the Mid-East headquarters of the International Center for Agricultural Research in Dry Areas (ICARDA), one of the United Nation's nine official seed banks and agricultural research institutions. Such centers are key to maintaining the earth's local seed resources in vastly different ecological settings. Among them are the International Center for Tropical Agriculture, with gene banks in Brazil and other tropical countries; the International Rice Research Institute, working to promote high-yield rice varieties throughout Asia; and CIMMYT, the International Maize and Wheat Improvement Center, headquartered in Mexico, a center of origin for many of our most familiar cereals and grains. These days it's ICARDA that's getting a lot of attention because so much of the earth is coming to resemble their specialty—"Dry Areas." What makes these seeds remarkable are the genetic stories they contain, evolving season after season, millennia after millennia, to shifting conditions, diseases and pests—just what we all need as the planet's food-growing lands start to look more and more like the Fertile Crescent.

"For a person like me, crossing the Euphrates River is very emotional because, you know, everything started there," recalled Salvatore Ceccarelli, a seed scientist who spent twenty-five years based in Aleppo as a senior breeder and researcher specializing in barley and wheat. He conducted several training courses for Iraqi scientists in Abu Ghraib, and helped to oversee the care of the Iraqi seeds in Syria after the center was destroyed. As of 2015, the Tal Hadya center had 153,000 samples, representing 680 crops, a collection rich also with wild land race relatives containing the parent DNA of today's food. The Syrian and Iraqi seeds, Ceccarelli told me, "could hold the secret to adaptation to drought . . . After ten thousand years more or less of evolution in a very dry place, we can see what natural selection has left behind."

During the Syrian civil war, Aleppo was a stronghold for the opposition. A group of armed rebels occupied the research station, but they didn't destroy it. Ceccarelli told me that they lucked out: some of the rebels themselves had been farmers before taking up arms against the Assad regime, and their commander had a degree in agronomy. He understood the significance of the seed facility, and "made sure throughout the battle in Aleppo that the center received enough diesel and gas to keep the power on and the refrigeration running." The rebels offered protection in return for allowing them to harvest the food growing in the center's fields. "We are not thugs," one of them guarding the facility told a reporter before the siege.[1] "We are decent people who fully understand the value of scientific research." By 2012, though, most non-Syrian scientists had fled. Those remaining, however, kept the seed bank functional and even succeeded in sending off a couple of shipments of seed duplicates to the Norwegian mountain vault in Svalbard.

Then in 2016 the fighting came directly and aggressively to Aleppo. The government dropped barrel bombs and assaulted the city with troops and tanks. The supply of electricity became erratic. If power was cut, the refrigerated vault would become overheated, which would be a disaster for the dormant seeds. In May that year, the remaining Syrian scientists scrambled together a van and loaded it with seeds, and took off on the lone open highway across the border to Lebanon. The seeds made it, with the acquiescence of Hezbollah, which controls the border region in Lebanon's Bakaa Valley. "You know you don't need a big truck for seeds, they're small," recalled Ahmed Amri, an ICARDA plant scientist who received the seeds on the other side of the border. "Nine boxes, six thousand accessions, on one small truck," he told me on a Skype call from Beirut. Small, but each one loaded with the potential to help us navigate the tumultuous times ahead.

Some of the most precious seeds on earth, containing priceless genetic history, a chromosomal map into how farmers might respond to the overheating climate, were on the run again. Like farmers and scientists before them had done in similarly disrupted circumstances, they saved what they could.

**THE FIRST KNOWN** instance of humans going to extraordinary lengths to defend their seeds dates back to 73 AD and the famous battle in what is now Israel at a hilltop redoubt in the Judea desert called Masada. After being besieged for months by Roman soldiers, the band of Jewish rebels in Masada famously opted to kill themselves rather than surrender. They left behind clay jars filled with date palm seeds—which they intended, according to biblical scholars, to signal to the invaders that it was not for want of food or provisions that they killed themselves, but a defiant refusal to be captured alive as Roman slaves. That incident has gone down in history as an inspiring act of defiance suggesting the potency of seeds as a symbol of human survival and sovereignty. (Within one hundred miles, the Palestine Heirloom Seed Library in the West Bank town of Battir, near Bethlehem, and the Israeli Gene Bank, just outside of Tel Aviv, preserve descendants of those seeds. By 2015, several of those two-thousand-year-old seeds, replanted on an Israeli kibbutz, were actually growing into fully developed date palms.)

In more recent times, a celebrated instance of "seed heroism" involved the seeds accumulated by the Russian botanist-explorer Nikolay Vavilov. The institute in Leningrad (now St. Petersburg) that Vavilov founded, which at the time housed one of the world's largest seed collections, faced a direct threat during the German Nazi siege between 1941 and 1943. Over the course of those two gruesome years, Vavilov's colleagues at the Scientific Research Institute of Plant Industry refused to succumb to the temptation of eating the seeds during a period of grueling hardship and starvation. Instead, they spirited them away, buried them, or tucked them into hiding places, to protect the seeds from destruction or confiscation by the Nazis—an effort made even more poignant by the fact that Vavilov himself was at the time imprisoned in Stalin's gulag, where he would die in 1942. "For me," Ceccarelli told me, reflecting on his visit some years ago to Vavilov's former office, "it was very emotional to see where he worked, all he did to find and save those seeds. It was very emotional, knowing the history of this great scientist."

In a poignant irony, during the chaotic dissembling of the Soviet Union in 1991, Ceccarelli and his wife Stefania, also an agronomist, were among a group of scientists invited to St. Petersburg and given several thousand accessions for safekeeping back in Syria, just in case the Institute's collection didn't survive. That fortunately did not happen—the Vavilov Institute's collection still stands today in St. Petersburg. Which means that some of the seeds stored in Syria and then sent off across the border into Lebanon are descendants of the seeds collected by Vavilov himself.

By 2017, the refugee seeds from Tal Hadya were successfully replanted in Lebanon's Bekaa Valley and at another ICARDA center outside of Rabat, Morocco, where Ahmed Amri lives. He commutes regularly now from Morocco to Beirut to oversee the plantings in the two plots, a process that is essential to keeping the seeds vital. They plant in November, he told me over Skype from Beirut, and harvest in March and April. "We're trying to reconstruct those Syrian and Iraqi varieties before they disappear," he said.

The Syrian seeds are already playing a significant role in responding to our storm of planetary climate disruptions—starting in the middle of one of America's prime wheat growing regions, in the state of Kansas. Degree by degree, Kansas is getting hotter. Between 2000 and 2015, the temperature in the Midwest rose from one to two degrees Fahrenheit above what had been the twentieth century average, according to the National Climatic Data Center.[2] That is a severely disruptive jump over a relatively short time span, and the consequences are already being felt. In Kansas and surrounding states—Oklahoma, Nebraska, and Missouri—the rising heat is triggering a surging population of Hessian fruit flies, which deposit their larvae in the leaves of wheat stalks. In the past, only a limited number of larvae would survive through the onset of winter; now, the cold comes later in the fall, which means far more larvae survive to attack the wheat. Farmers are seeing an average 10 percent yield loss due to that one pest alone—a big loss for farmers already operating on narrow margins, according to scientists at the Wheat Genetic Resource Center at Kansas State University (KSU) in Manhattan, Kansas.

Ahmed Amri, who received his PhD more than two decades ago from Kansas State, is in touch with his alma mater: he's helped to arrange shipments of Syrian seeds to KSU's Wheat Genetics Resource Center, which orchestrated a simple Darwinian experiment. They planted the seeds of a wild wheat relative, common in the Mideast, in a greenhouse alongside common Midwestern varieties and let them grow into seedlings for about three weeks. Then they unleashed the fruit flies into the greenhouses and checked to see which survived. It turns out the seeds from Syria are, by far, the most resistant—showing at least 25 percent more resistance than the commercial varieties planted in the American wheat belt. "That's because they're adapted to a Fertile Crescent climate that is coming to resemble the climate in Kansas," explained Ming-shun Chen, an entomologist and professor of molecular biology at the university who helped design the trial. "Those are good seeds (for us)," he told me. "And they don't need any pesticides to kill this insect." Those refugee seeds will be bred into the existing seed populations, and are expected to play a key role in saving the American wheat harvest from disruptions being wrought by climate change.

Further north in Illinois and North Dakota, there's another symptom of climate chaos for which Syrian seeds appear to be the cure: here an aggressive fungus, called *Fusarium*, is flourishing in the moist fields left behind by rainstorms during a heat wave. "We've had a failure with the modern wheat varieties lately due, I think, to the hotter temperatures from climate change combined with these crazy rain events," said Bill Davison, an ag extension agent with the University of Illinois,

Bloomington. Davison told me that this has impacted not only the quantity of grain produced per acre, but also the quality—pushing the grain that is produced down from the highest grade for human consumption to the lowest, animal feed grade, a financially devastating demotion. After numerous experimental trials he's also finding that seeds evolved from Syrian parent lines—a strain he calls "Banatka"—are showing high rates of resistance to the fungus. In "normal" conditions, he said, their yield is lower than the industrial commercial varieties. But, said Davison, there no longer is a "normal," or at least the one they'd been accustomed to. The Syrian seeds perform much better under the conditions of the new "normal," and the Fusarium fungi that come along with it. "What we've been seeing with these consecutive years of rainy hot summer is that these (Syrian) varieties have done very well," he told me. "In these conditions they at least make food grade, because they have resistance to the Fusarium fungus."

Seeds travel easily across frontiers, taking with them the qualities that may be useful in new locations experiencing similar conditions as their home terrain. Like Chandler, Arizona, for example, in the field of Hussein al Hamka, an immigrant from Iraq. Al Hamka lived for many years in the Tigris-Euphrates valley—in the heart of the Fertile Crescent—on a seventy-five-acre farm outside of Mosul, Iraq until 2007. Back in the nineties, he told me, he started planting wheat and bean seeds that he obtained from the seed bank in Abu Ghraib. For him, it was just the local seed. Then, in 2007, al Hamka reached his limit of Iraq's cycles of violence, made worse by the discrimination he and his wife experienced as ethnic Yazidi. They fled the violence across the border to Syria, following a similar route taken by the Abu Ghraib seeds themselves four years earlier.

The couple spent two years in a Syrian refugee camp. In 2009, they obtained visas to come to the United States. The two flew to the

United States and, with the assistance of the International Rescue Committee, were steered toward resettlement in Arizona. Within a year, the Committee had helped them obtain access to a small farm in the town of Chandler, outside Phoenix. When I spoke with al Hamka by phone in 2017, he told me that every step of the way in their long journey to America he carried a pouch of seeds from his farm, for five different crops—less than half a pound, he said, but that was enough. "For me, they had the same value as gold. It was like keeping our family together, keeping our seeds," he said. He brought those seeds with him to Arizona.

In a landscape quite similar to the one back in Iraq, he's been growing cucumbers, cantaloupe, chard, and radishes based on seeds they brought from home. And his Iraqi durum wheat became a prized commodity among heritage grain aficionados in the Southwest. Unlike the commercial strains we're accustomed to, the Iraqi durum has frilly awns that are long and black, not "golden brown" like the typical American wheat field. The seeds are pale yellow, and they perform well in the arid climate of the Southwest, which is a lot like Iraq, and far more resistant to water stress than the hybrid commercial strains grown on large, chemically dependent farms in other parts of the state. His wheat and other crops are popular at the Downtown Phoenix Farmers Market, and with a local family-run flour mill, Hayden Flour Mills, renowned for its stone-ground heritage grains and cereals. "It's more nutty and a little sweeter than the commercial varieties," commented Emma Zimmerman, who runs the mill with her father. Brought to America by a refugee, the seeds are bringing not only new tastes, but new vigor, without pesticides or genetic engineering, to the struggle against advancing heat and retreating water.

The overall global threat to such precious seed resources is for the most part, of course, quite different than the threats faced by seed banks under siege in Syria or Iraq. The world's seeds are not so much threatened by explicit conflict, as they are by the conflict over clashing visions of how we grow our food.

The seed movement itself is growing concentrically outward as farmers, scientists, native peoples, and health-minded consumers find themselves in league in an effort to sustain diversity and the ecological balancing act that a healthy food system depends upon. Just as those farmers in Syria and Iraq were focused on sustaining their rich pool of genetic resources, a rapidly growing popular movement across the United States and around the world is asserting the right to choose seeds embedded with the germplasm that has evolved and survived through generations of shifting conditions in one place, and given life through techniques that work with, rather than against, the conditions in which they're grown. Though in drastically different circumstances, they're motivated by the same impulses as farmers in Abu Ghraib and Aleppo, and all the other myriad of places where seeds are a way of retaining historical memory embedded in germplasm, steeped in the rich soils of people's land and culture.

Our genetic heritage is caught between the pincers of climate disruption and corporate consolidation. That van racing across the Lebanese frontier seems an apt image for what is becoming a global race against time to preserve the world's seeds.

**IN TUCSON, ON** North Stone Street, just off Broadway in the heart of downtown, lies the main branch of the Pima County Public Library. Within sight of the checkout counter, just to the right of the glass entrance doors, a couple of old weathered wooden catalog boxes, salvaged from the storeroom after the library went digital, are piled on top of each other, nostalgic relics of 1970s-era library décor. Here is the Seed Library within the city's library of books, the largest publicly accessible collection of seeds in the American Southwest.

I pulled open one of the long rectangular drawers. Inside are not the names of books and authors on little index cards, but seeds, hundreds of them, tucked into little plastic packets. These seeds have plenty of stories to tell, like the books lined up neatly on the shelves. Except seed stories are told in the non-fiction language of chromosomes.

Justine Hernandez, a librarian who worked with community members, farmers, gardeners, and the local city government to launch this library of living things within the library of books, showed me the collection, opening and closing the drawers one after the other. There are seeds in every shape and color—black, brown, speckled yellow, dark red, dull white—all of them adapted to the specific hot and dry conditions of the Southwest, including numerous variations of squash, corn, beans, chilies, cucumber, and more. "Each season," Hernandez said as we flipped through the seed drawers, "we're building up a repository of seeds that survive as the temperature goes up and we see less and less rain. We're building adaptability, seeds able to survive in these changing conditions."

As one local seed company after another were snapped up by chemical companies, the libraries are part of a growing parallel movement that's been pulling in the opposite direction, away from homogenization and toward diversifying local seed stocks and strengthening the role of farmers, not companies, in sustaining them. Since 2010, more than four hundred of such libraries, of various sizes, have opened across the United States. They're now in every region of the country, one of the multiple edges of a rapidly growing movement. "In a world in which more than half of our seeds are controlled by three companies, we need to re-localize our resources," said Rebecca

Newburn, director of the seed library in Richmond, California. Similarly installed in a corner of the main public library, that one focuses on seeds for the relatively moist, fog-heavy eco-pocket of the San Francisco Bay Area, and represents future crops of kale, peas, bok choy, chard, tomatoes, and many more.

The same principles govern the seeds as govern the books: farmers or gardeners can "check out" seeds, plant them, and are expected to come back the next season with another round of seeds—along with comments on how they perform in the local conditions. All focus on varieties adapted to local conditions, and on sharing the knowledge that farmers and gardeners develop in cultivating them.

Of course, creating an independent space for seeds and seed testing independent of the conglomerate breeders was bound to collide with the powerful regimes that govern the seed trade. On June 12, 2014, the collision occurred. A seed library inside the public library in the western Pennsylvania town of Mechanicsburg received a notice from the Bureau of Plant Industry, part of the state Department of Agriculture. The language in the letter, a copy of which I obtained, conveyed a bureaucracy grappling with the muddy imprecise world of independent seeds. "It has come to my attention that you intend to offer your patrons the option to participate in a seed library," said the letter, signed by the state's Seed Program Supervisor for the Bureau of Plant Industry. "My understanding is that patrons will be able to 'check out' seeds, take them home and plant them, harvest any resulting fruits, collect seeds, and return the collected seed to the 'Seed Library' for planting in the following season . . . I believe there are some issues of seed distribution that you may not be aware of . . ." The letter then went on to list several statutes that the library could be violating, including the dissemination of unregistered and untested seeds, and seeds that had not been officially assessed for their germination capacities.

Putting a library on notice that it could be violating a law for freely disseminating seeds almost stopped the budding movement in its tracks. But to the librarians' surprise, the story went viral—or, as Rebecca Newburn put it, "it went fungal." Soon stories were emerging in the regional media about the state's effort to close a *library* that was disseminating seeds. By 2015, a small NGO in Oakland got involved, the Sustainable Economies Law Center, which promotes legal reforms to encourage locally based economic initiatives.

Neil Thapar, staff attorney at the Center, said that the Pennsylvania threat was a wake-up call. The seed libraries were indeed in a legal gray zone. They weren't quite legal: they were disseminating seeds not authorized or tested by the state, which does so to protect consumers from seeds that might deliver a different plant than promised. But they're also not quite illegal, since they were not distributing patented varieties, which had gotten farmers in trouble in the past.

Thapar and his colleagues went to work building coalitions with local farm and environmental groups in Pennsylvania, and convinced the newly elected Democratic governor to retreat from the aggressive stance taken by his Republican predecessor. Over the next two years they moved onto other states that had thriving seed library networks threatened by regulators. In Minnesota, Nebraska, Illinois, and California they succeeded in getting laws enacted that clearly affirm the rights of libraries to disseminate seeds and exempt them from the regulatory regime that applies to commercial seed companies.[3]

The libraries have come to symbolize resistance to the steamroller of consolidations that have transformed the seed industry over the past twenty years. We've evolved so far down the patented seed path that the simple act of planting, saving, studying, and breeding seeds is now seen as a fundamental act of defiance to the handful of large agro-chemical companies that dominate the food system. One of the most consequential impacts of the seed library victories has been to reaffirm the principle, lost in the haze of consolidations and patent extensions, that farmers' knowledge and experience is central to the breeding process. "Farmers are not 'consumers' requiring 'consumer protection' from seeds," Thapar said. "They are active and engaged participants in sharing, testing, and using their seeds." Each of those seeds contains a story of evolution, a history, written in genetic form, of a plant's call and response with the elements as it struggles to survive in shifting conditions.

**TO EXPAND DIVERSITY** and the possibilities for new combinations, farmer-run seed exchanges are blossoming across the country and around the world. In the United States, from Iowa to Idaho, Tucson to St. Petersburg, Florida, and Highland, Texas to the Hudson River Valley of New York, seed exchanges offer multiple options for farmers to choose from, purchase, test, and return the next year to "exchange" for another round. Native seed exchanges—so named because they're seeds that have evolved over many seasons in their home ecological pockets—are proliferating across the country, like the one in Tucson which had garnered the unwanted attention of Frito Lay's cease and desist letter. These are centers for breeding, studying, and selling seeds adapted to their native ecological niche. They're similar to the libraries but also include testing in experimental plots, and frequently offer seeds for sale.

They're like dating services for seeds—offering abundant options for throwing together new combinations in a field to see what sticks and what doesn't. In the middle of the Midwest, in Decorah, Iowa, the biggest of them all unfolds in July every year when the Seed Exchange hosts the breeders of multiple varieties of cereal, grain, vegetable, and fruit varieties in what's become a global seed bazaar.

In the Northern Rockies, Bill McDorman, director of the Rocky Mountain Seed Alliance (and former director of Native Seeds/SEARCH) is working with farmers in the mountain states to seek and breed out native potatoes, grains, and cereals that have practically disappeared under the pressure of large scale commodity farmers. "Seed libraries and exchanges may be the most important thing we're doing for agriculture in the twenty-first century," McDorman said.

A parallel effort is underway to create an alternative to plant patents, a kind of creative commons for seeds to return us to what's actually a pretty simple idea: farmers testing and experimenting with what works in their fields in the particular conditions they face. The Open Seed Source Initiative offers new varieties for breeding and for sale that are explicitly not patented. Farmers, scientists, or breeders using them have only to agree to stipulations that they will not patent the seeds developed from the parent lines, or otherwise limit their use—and in return are free to use them as they wish, for crops or breeding stocks or research. Jack Kloppenberg, a longtime professor of Agricultural Economics at the University of Wisconsin-Madison, who launched the Initiative along with a group of fellow agricultural scientists, activists, and farmers, explained that the idea was to find a way to protect farmers' innovative creations in the field while not limiting access to them by other farmers. "We're taking our inspiration from open-source software," he explained, "and applying those principles to seeds." So far, about fifty crops, from amaranth to zucchini, are available under the open seed pledge. "You can't have sovereignty in food," Kloppenberg said, "if you don't have sovereignty in seeds."

* * *

**WHEN HUMANS FIGURED** out how to predictably wrest edible foods from the earth and our wandering in search of nourishment stopped, the particular ways in which we domesticated plants became central to our multifarious cultures and identity. How we grow food, what we eat, and how we distribute the fruits of our harvests became central to our sense of place in the world. Food and the rituals around it course through our cultural, religious, tribal, and national identities.

I'm sure I'm not the first person to say that meals in a traditional Jewish household are a far different experience than eating in a, say, traditional Protestant, Catholic, Muslim, or Buddhist household. Or in a traditional Navajo, Sioux, Colombian, French, Mexican, Ethiopian, or Chinese household. Or a North American Caucasian or African American household.

"The rituals, not to mention the recipes, that occur around these different tables all have their historical roots back in the soil," commented Alice Waters, the legendary pioneer of food consciousness, as we walked through the garden she created at a middle school near her restaurant, Chez Panisse, in Berkeley, California. Waters's aim when she founded the restaurant three decades ago was to celebrate locally grown food. She's spent the last two decades working to expand those principles into the nation's schools, using school gardens and accompanying curriculum to teach the fundamental principles of farming and, of equal importance for those not destined to land on a farm, the politics and economics and science of what lies behind growing healthy food. The garden at Martin Luther King middle school was her first Edible Schoolyard; I used to run on the school's track (on weekends) and would take a quick diversion through the pathways of chard and kale, broccoli and peppers as teenagers of multiple backgrounds seemingly happily dug trenches and sprinkled the crops with water. Waters famously teamed up with Michelle Obama to create the White House's first organic garden. Her curriculum has been adapted in one form or another by more than five thousand schools in the United States and overseas. It starts, she said, with learning how to plant seeds. Then it expands into learning what those seeds tell us about, "the story of a place, of how the seed got there, of the people who cultivate and consume the foods that the seeds deliver to us."

Such stories, re-connecting seeds back to the people who grow them, are moving up the food chain. In 2016, the London-based Gourmand Society, an association of the world's top chefs, gave its top award for best cookbook to a six-hundred page, six-pound tome in three languages—English, Tajik, and Pashtun—called *With Our Own Hands*.[4] It was written not by chefs, but by a Dutch ethnobiologist, Frederik van Oudenhoven, who lives in Amsterdam, and a Swedish grad student in anthropology and agricultural development, Jamila Haider. She lives in Stockholm and works for the Stockholm Resilience Center, a think-tank which promotes strategies for adapting to environmental stress, most notably climate change.

The two spent two years researching the people who live in the Pamir Mountains of Central Asia, one of the most bio-diverse and remote mountain ranges, straddling Tajikistan and Afghanistan. The Pamir was the first region that the Russian botanist Nikolay Vavilov visited in his lifelong search for seeds; here he found cold-resistant and early maturing cereal and grain varieties that he sent back to Russia for planting in the country's cold north. Altogether, an incredible forty-seven different varieties of wheat have been identified in these mountains.[5] In this harsh, beautiful landscape, Haider and van Oudenhoven collected stories from the people who live in the mountain communities, and observed the relationship between their foods and their culture.

You'd have a hard time cooking many of the concoctions, particularly if you don't have access to ingredients like yak milk and mulberries. It's really an anthropology book with recipes, filled with stories of people's lives, their homes, and their fields. For the first time, a book was honored by this group of esteemed chefs not necessarily for the specific recipes, but for the authors' extraordinary ability to weave the cooking traditions with the practices of cultivating, harvesting, and cooking food. The Gourmand Society praised it for "revolutionizing" the very concept of a cookbook.

Haider told me that the genesis to the "cookbook" part of their anthropology project came out of a simple question: after trekking and researching for months across the region's spectacular peaks and lush valleys, it dawned on her and her co-author that in a mountain range with multiple varieties of wheat, they were always served one kind of white bread in the village restaurants and roadside stands, along with greasy imports often from Russia. Where was all the diverse food in this biologically diverse region? They started asking, and many of the elder women started coming to them with their traditional recipes, which they feared would be lost as their children moved away to the cities. The women started cooking the local dishes for the two visitors, and in turn the two started asking questions about the dishes. The answers they received became the book, which uses the stories of food and farming, and the recipes that go with them, to help tell the story of people in one of the more inaccessible places on earth.

No one was more surprised by the award than the two authors. "We didn't set out to write a cookbook," Haider told me. "Then we heard that its 'revolutionizing' the way a cook book could be. It's not about the food *per se*. We set out to document recipes as a way to unlock memories, the history of a place."

You could do that kind of research in any corner of the world. Genetic resources are far more than an assemblage of genes on a helix, but are the source of all that connects us to this patch of territory we inhabit, wherever it may be. Every seed is both a simple pocketful of genes, and a multi-dimensional and complex "packetful" of stories, like this book. Unpack them and you get the history of conditions below and above the earth, you get traditions and identities that have grown up alongside the foods of a place, and you get a portrait of the people who cultivate and nourish the foods they eat. You get a sense of the power that comes with controlling access to them. And increasingly you get a whiff of the struggle underway to resist that power, hold onto those traditions, to maintain that connection to our food, and to assert a different kind of power.

# CHAPTER 7

# POSTCARDS FROM THE PARADIGM SHIFT

**WHAT FOLLOWS ARE** three stories, set in three very different locales, three portraits of confrontation with the industrial food paradigm.

## BECOMING ARIZONA

I took a trip into the future of agriculture, and landed in Arizona.

When I arrived in Tucson, it was another 107-degree sizzler in a record-breaking spree of hot July afternoons. The heat was dry and seemed to suck the moisture out of everything. A water crisis was rattling nerves; a local radio station played an entire segment of songs devoted to the topic of "rain." They weren't playing pleasant songs like "Raindrops Keep Falling on My Head," but anguished calls for more of it from the likes of Tom Waits, David Crosby, and U2.

Driving through the vast stretches between Phoenix and the Mexican border that summer in 2016, it was mile after mile of shrubs, quirkily shaped cacti, and slithering lizards. And then, suddenly, a field, low, green, velvet in the bristling sun—row after row of lettuce, chilies, melons, and cucumber, appearing like life preservers bobbing in the dead calm Sonora desert. Arizona is an unlikely three-billion-dollar agricultural powerhouse, the leading producer of lettuce and one of the top producers of winter vegetables in the United States. In the central part of the state, you can smell the crops before you see them, a mist of nitrogen fertilizers, acrid pesticides, and moisture from the water delivered by irrigation canals to the desert.

Here the impacts of climate change are being felt acutely. The state boasts one of the nation's first TV weather forecasters, on the Phoenix ABC affiliate, to regularly connect climate science with her weather reports. The number of days above 100°F is rising quicker in Arizona than practically anywhere in North America, and the amount of rain plunging to record-breaking lows.[1]

Arizona won't be alone for long. Bands of rising heat and declining water in climate models by the National Oceanic and Atmospheric Administration (NOAA) are moving steadily northward.

George Frisvold, an agricultural resource economist at the University of Arizona in Tucson, described the shifting boundaries in the United States. "Ohio will look like Kentucky," he told me. "Kentucky will look like Tennessee. Tennessee will look like Alabama. Central California will look like Arizona." Conditions that have long reigned in the forbiddingly dry northern Mexican state of Sonora are coming to Arizona and New Mexico. Some of the richest agricultural areas on the planet, in places like South Africa, Australia, and southwest China can look to Arizona for a glimpse into their own future. North Africa, much of the Middle East, and large food producing areas of the United States, including parts of California and Texas, are already there.

The fields of Arizona may look green but keeping them that way requires the most complicated plumbing system in the United States—the dams, reservoirs, and canals which channel water from the Colorado River. Shuttling water from mountain wet zones to nourish farms in dry zones—a model typical also in California and other food growing regions—is being rapidly outpaced by changing precipitation patterns and diminishing river flows. Entire water networks, and the political networks that grew up around them, are being upended by the realization that there will no longer be water without end.

Politically conservative, ecologically dry, and in the climate bullseye, Arizona is one of several emerging hubs in a global seed movement that is slowly but steadily transforming how we view the industrial agriculture giants and their influence on the food system. By the luck of geography, or lack of it, the state is something of a test laboratory to assess our seed options as predictable patterns of weather and water dissipate. It's at the front line, among numerous front-lines, of the fight to preserve the evolutionary bounty of the earth's seeds in the face of unprecedented economic and ecological pressures. The key difference between Arizona and the many other centers of this growing movement in the United States and around the world is that there the hot conditions predicted for much of the world have quite definitively arrived.

So after a two-hour drive south from Tucson, I pulled off highway 19, drove up a meandering dirt road for half a mile, and landed at the doorstep of Gary Paul Nabhan, one of the world's leading ethno-botanists and an acclaimed expert in desert agriculture.

**NABHAN IS A** towering figure among those who plant or study seeds in ecologically stressed environments—which is to say, precisely the conditions we now face. His knowledge of modern and ancient farming techniques, developed over thirty years of studying and growing food in the desert, is ideally suited to this time of ecological disruption.

I met Nabhan at his home perched on a bluff in the high desert of southern Arizona, where he lives with his wife about fifteen miles north of the Mexican border. From Nabhan's living room, we looked through a picture glass window at his startlingly green, terraced fields, set amidst the hotly lit brown chaparral. An olive orchard was tucked into a valley in the distance. About five miles downhill is the town of Patagonia, where the authors Jim Harrison and Cormac McCarthy had homes; both wove the area's forbiddingly stark beauty into celebrated novels. The landscape can look barren from afar, but after a while you notice it teems with life, the lizards flickering along the desert floor, blossoms unfolding on the cacti, the birds overhead and the strange buzzy sound of small mammals and a million insects.

Nabhan's working his forty acres on a multi-million-dollar USDA grant to test which crops perform best in dry, hot conditions. His miniature apple, pomegranate, pear, and manzanita trees were terraced down the hillside; at ground level, lettuce, chilies, and tomatoes proliferated under their shade, along with desert herbs like sage, rosemary, and basil—a desert vista of Vavilovian diversity if ever there was one.

We walked outside. There hadn't been rain for weeks, so the soil at Nabhan's place was dry, and the leaves seemed to be retreating. But it was also fecund—the earth smelled of, well, earth, and it was colorful, with tiny flowers popping from stems or peeking above the surface, and the place buzzed with insects and occasional birds, an oasis of cultivated agriculture with no irrigated source of water. The terraces we walked on were based on designs from neighboring native communities, with whom he's collaborated, cut angularly into the hillside with small trenches to ensure as much water storage as possible when it rains. All were planted with seeds that are indigenous to the desert, which means they are deeply rooted to access underground water sources; none come from the big seed companies, which dominate the irrigation-dependent, industrial farms to the north, and which rely on pesticides and other chemical biocides, with the associated public health impacts in surrounding communities. Altogether, Nabhan grows about 150 different fruits and vegetables. "That variety allows me to hedge my bets," he said.

Squat, with a boxer's build, Nabhan is accustomed to working outside in overalls, but capable of putting on a suit and tie when necessary—occasions which are occurring more frequently with political figures in his state and internationally who are worried about food security as the pace of climate change accelerates. The weather on his farm is behaving just like model after model predicts. When I visited him that July he told me that he and his wife and other farmers in the area had been preparing on and off for weeks for one of the periodic monsoons that drench the desert. But the signaling, like so much else, is off-kilter: the weather oracles looked repeatedly to the distant dark clouds and predicted the summer storms were heading their way, and time after time they did not come. Monsoons are what supply much of the state with whatever water it gets independent of the Colorado River, and are the primary source for the replenishment of underground aquifers. But Arizona has been receiving only about two-thirds of the rainfall that it was accustomed to before 2010.

Nabhan retraced the steps taken by the Russian botanist-explorer Nikolay Vavilov in his book *Where Our Food Comes From*. As the range of desert-like conditions expands, Nabhan said, Vavilov's relevance expands with them. We passed alongside a row of miniature apple trees. "The seeds Vavilov encountered in every desert around the world have even greater significance to farmers everywhere today than they did when he first made his explorations . . . What we do here in this little laboratory has ripple effects into all those areas facing the desert for the first time."

Until recently, Nabhan's work studying and growing food in desert environments was considered a fringe endeavor. Through most of the twentieth century, centers for the study of desert agriculture were shut down or severely cut back at land grant universities across the United States; it was presumed that there would always be enough water. It was not considered relevant to learn from those who mastered the art of growing food in places where the water does not fall predictably from the sky. That was the assumption in Arizona, which plumbed the Colorado River; just as it was in neighboring California, which plumbed the Sacramento and San Joaquin Rivers to create the Central Valley, the powerhouse of industrial-scale fruit and vegetable crops; as it was in the Midwest, which sent wells deep below the earth's surface to feed the massive cereal and grain farms with water from the Ogallala Aquifer. All three of those food centers have experienced record-breaking temperatures and droughts since 2010.

Nabhan is at the forefront of efforts to respond to these changing conditions by returning to basic principles—looking not to company laboratories for the most successful crops, but back to Native Americans. Arizona and the Southwest has the longest history of domesticated agriculture in the United States, dating back at least three-thousand years. The state was carved out of the desert by miners and ranchers who violently displaced tens of thousands of Native Americans, who'd built societal networks centered around agriculture. Many of their descendants remain, however, and are helping to teach us how to navigate the earth's disturbed landscape.

The dryland seeds nurtured in this region over millennia have adapted over many generations to the searing heat of summer, periodically biting cold of winter, and scarce water practically year-round. The Tesuque Pueblo north of Santa Fe, in New Mexico, has been one of the centers of this resurrection of native agriculture. The native Tewa community has been using a combination of Western scientific and native knowledge to grow corn, beans, squash, and other vegetables to provide free food to the community. "Seeds are the first link in the food chain and this link is now under threat," Clayton Brascoupe, a Mohawk/Anishinaabeg farmer told a Mountain West Seed Summit, a gathering of local tribes in March 2017. "Our responsibility is to preserve them for future generations." On a global scale, the Indigenous Environmental Network, a global association of indigenous communities and tribal governments, has made seed saving a priority—and many of us non-native people may end up grateful they have, as interest rises in the seeds that survived over many years of changing conditions.

In Arizona, Nabhan works with First Nation Navajo, Apache, Hopi, and Tohono O'odham, who have farmed the lands of southern Arizona and New Mexico for many generations. They offer access to seeds and accumulated knowledge of how to grow them, for which Nabhan offers insights from the modern agricultural sciences—the latest in non-chemical pest and disease control, for example. Any new seeds based on native seed parent lines that come out of Nabhan's trials are provided for free to native communities. Nabhan is something of a bridge between Western scientific and native peoples' experience-based botanical knowledge. The latter, of course, has been hiding in plain sight in the United States for more than two centuries.

**IN 1930, NIKOLAY** Vavilov made his way to the University of Tucson, then an outlying desert town in Arizona, which had become a state just fourteen years earlier. He lectured to a packed house at the University of Arizona Library Auditorium, where he outlined his ideas on the links between plant breeding and biodiversity, and then spent several days visiting with local farmers and agriculture scientists. The Russian, unlike most American agriculturists at the time, sought out leaders of the Hopi, Navaho, and Pascua Yaqui tribes who had inherited practices built over generations of food self-sufficiency in the desert. Vavilov traveled to the Hopi and Navajo reservations, and observed with awe their abundant fields of corn, beans, squash, amaranths, sunflowers, wild rice, and Jerusalem artichokes even as Dust Bowl droughts were devastating farms in other parts of the Southwest.[2] They shared knowledge, Nabhan told me, about the crops best suited to drought resistance, terracing methods for efficiently channeling rain, and planting techniques for maximizing water absorption in the soil.

Down the road from Nabhan's place is the farm he founded to put such revelations into practice, the Native Seeds Conservation Farm. The farm spreads over sixty-five acres alongside a narrow, rock-filled gully that is a key route for migrants heading north from across the border. Here the seeds collected at the Native Seeds/SEARCH vault in Tucson are kept vital, tested out for performance, and multiplied. Seeds are provided free of charge to Native American farmers in the region, for whom the farm offers basic farm training and field experience. At the edge of a newly planted row of beans, I met Robyn Pailzote, a twenty-year-old member of the White Mountain Apache. She's working here along with other young native farmers—from the Navajo and Tohono O'odham—to test the performance of seeds for crops that have been grown in this area for centuries, including squash, peas, and corn, arrayed in even furrows, seedlings perched just above the surface. She told me that her grandparents were dry desert farmers, and she wants to continue that tradition.

When I caught up with Pailzote she was planting a row of tepary beans—a tasty legume that has been cultivated in this region since long before Europeans showed up and is today gaining popularity across Arizona and the Southwest. She told me she's been eating the beans her whole life, and was happy to be learning how to grow them. I picked up a pound to bring home, and tried them for the first time; simmered with chilies and spices, they're meaty and slightly sweet, smaller than the usual commercial beans, but tastier, with an earthy crunch.

Off in the far end of the field is Sonora white wheat, a grain that evolved in the Sonora desert, sold and traded locally among tribal communities and across what is now the US-Mexico border for centuries. It's nuttier than your usual processed white wheat, and has found a thriving market throughout the Southwest, including at some of the hippest restaurants and micro-breweries in Tucson and Phoenix. It doesn't take long to figure out that some of the tastiest foods in the Southwest these days can trace their roots back multiple centuries—to plants that, like Indian parched corn, have adapted to the very conditions that major food growing regions of the world now face. They also provide an income stream to native farmers, selling to a budding constellation of heritage food restaurants in the region. The flourishing of indigenous foods in the Southwest was even recognized by UNESCO, which in 2016 anointed Tucson the first World City of Gastronomy.

**IN TUCSON, I** walked into the giant refrigerator back at Native Seeds/ SEARCH headquarters, the place to which that letter from Frito Lay about Corn Chips™ had been addressed. The cold air was refreshing on another overheated afternoon. Inside were hundreds of samples of

Southwest seeds. I stopped at a canister of ragged-edged oblong seeds, a jumble of dark browns and blacks—*teosinte*, "the grandmother of all corn," said Laura Jones, the organization's acting director. "Grandmother" meaning the fount of the genetic material contained in the corn planted across the world, which helped to save the American corn crop when it was devastated decades earlier by disease due to genetic uniformity. On another shelf: bean seeds, peas, and corn, in oranges, browns, and yellows. Then there was the barrel full of Sonora wheat seeds, yellow and white. I grabbed a fistful, hard and cold; they rippled through my fingers, chunky bundles awaiting sun and water. Recently, they've been contacted from places as far afield as Kenya and Chad since the climatic conditions in many key food producing areas of Africa—and other continents—are coming to resemble the hot and dry climate of Arizona, or, equally likely, the reverse.

Arizona is one small part of the surging interest in foods that have stood the tests of multiple seasons with multiple variations in conditions, many of them cultivated by native communities across the United States—wild rice in the Dakotas, indigenous corn, squash, and beans in Arizona and New Mexico, heritage grains like amaranth and kamut in the Northwest. When Slow Food Nations held a three-day festival to celebrate native foods in Denver in July 2017—Alice Waters helped organize the menu—many of the dishes served were born from these crops and their cousins.

As I visited the native seed farm that summer, about a hundred miles north, Monsanto was closing a deal to purchase 155 acres in Yuma county, the center of big-ag in the state. The company's stated goal: to pursue experimental trials of seeds genetically engineered to withstand heat and drought. Here, the company announced, they would build a thirty-foot-tall greenhouse and develop a "key hub for (corn) seed production and innovation." By early 2017 the plan had prompted major public protests. Under pressure, the company dropped its demand for a property tax waiver after local activists pointed out that the major beneficiary of the taxes it was trying not to pay were local school districts. Monsanto went ahead with its research and breeding center anyway, with the aim of devising a new generation of what's become known as "climate friendly" crops.

There was an eerie asymmetry: down south, native seeds with genes that reflect their survival over years of give and take with drought and monsoon, the shifting angles of climate, water, and pests; and up north, in a greenhouse, all that call and response short-circuited with inserted genes.

But as exterior conditions shift and become increasingly difficult to control or predict, an accumulating body of scientific evidence and farmer experience is challenging the idea that adaptive survival traits can be replaced by chemicals and genetic engineering. Climate change is putting industrial agriculture on trial. In the process, it's forcing us to confront the fundamental question: who do we want to rely upon for our seeds as growing food becomes more perilous and uncertain?

## STORMING THE CATHEDRAL

Hans Herren looks dapper now, with professorial spectacles and a mane of swept back grey hair. It's a long way from the muddy soils of Kenya and East Africa, where Herren made his reputation. There, the entomologist devised non-toxic ways of controlling insects and ecologically sound ways of growing food—for which he won the World Food Prize, agriculture's equivalent of the Nobel. We met on a foray he made to San Francisco from his farm in the Capay Valley, where he and his wife, also an agronomist, now grow organic vegetables and fruits.

When the World Bank approached Herren in 2003 to head up a research project into the most productive ways to grow food in developing countries, he was one of the most high-profile and globally respected figures in international agriculture. So it was a surprise to his sponsors when he and his team would go on to challenge the agricultural practices the Bank was promoting.

The World Bank's initiative was triggered in response to a highly effective pressure campaign by a coalition of environmental and food NGOs including the Pesticide Action Network, Friends of the Earth, and Food First calling out the bank's placement of agri-chemicals and genetically engineered seeds at the core of its multi-billion-dollar agricultural development strategy. Under mounting pressure, and to get the groups off its back, in 2004 the Bank agreed to pay for a study of its strategy, and for maximum credibility committed to finding an independent scientist to produce it. They tapped Herren, who at the time was running the International Center of Insect Physiology and Ecology in Nairobi, to run it. The Food and Agriculture Organization and fifty-six governments, including the United States, agreed to pitch in and commit resources to the endeavor. Herren now refers to the Bank and the FAO and the other major international agricultural development institutions, and the big-ag interests they have often represented, as "the cathedral."

Herren corralled a team of more than four hundred experts—recruited from academia, aid organizations, agriculture industries, farmers, and NGOs—who fanned out to food growing regions in every corner of the earth. The Bank hoped that this array of firepower would finally demonstrate the validity and importance of new agricultural technologies for addressing food insecurity. It would be the most comprehensive study of industrial agriculture ever conducted.

After four years, the verdict came in 2008, in a two hundred plus page report with a clunky title, the International Assessment of Agricultural Knowledge, Science and Technology for Development (IAASTD). Their conclusions, in a nutshell, amounted to a frontal assault on the mainstream agriculture consensus: the chemically driven model that had been central to "Green Revolution" strategies in developing countries, and from which agri-chemical companies drew their legitimacy in developed countries, were not delivering their promise of sustainable or equitable development. Modern industrial hybrids and GMOs, they found, often put small farmers into financial spirals, with expenses rapidly outpacing their ability to pay for them when the subsidies for the hybrid seeds and chemicals ran out. Rather than enhancing webs of food production and distribution, global trade deals disrupted them and often undermined local food production. One result of the NAFTA trade accord, for example, was that Mexico was flooded with cheap American corn exports, which led to the displacement of thousands of Mexican corn farmers who could not compete with cheap American corn. Many of those farmers ended up migrating north across the Rio Grande—and landing on the other side as targets of prejudice in one of the most divisive issues in the United States.

The approach Herren's commission recommended is called "agricultural-ecology," or ag-ecology, which involves treating a farm very much like holistic doctors treat the innate metabolic capacities of humans to defend ourselves from infection and disease, and to bounce back when they do strike. That means channeling the feedback loops of a healthy and diverse ecosystem. Such practices rely on natural methods for repelling threats and regenerating soils, including cover crops to return minerals to depleted soils, encouraging the presence of beneficial insects to control pests, rotating crops to keep pests off balance and working more organic material into the soil to absorb water and provide a rich nutrient base. These techniques, they argued, are also far more resilient to changing conditions than the industrial, single-crop farms reliant on agri-chemicals for sustenance.

In countries facing severe environmental stress—Burkina Faso and Ghana, in West Africa, and Haiti in the Caribbean, for example—researchers affiliated with Food First found multiple examples of farmers using these strategies had success in dealing with drought and disease while conventional varieties failed.[3]

Here was an endorsement by leading figures of agricultural development that mixed the insights of modern agricultural science with indigenous knowledge based over generations of farming in specific ecosystems. It also recommended including more farmer input into seed and cultivation strategies; less of one-size-fits-all technological inputs; and limiting the use of agri-chemicals to highly targeted and discrete purposes, as opposed to the blanket spraying of fields that has become routine in developing—as well as in developed—countries. All of which, in sum, challenged the long-reigning assumption that single-crop, large-scale farms being promoted by international aid institutions—like, indeed, the World Bank itself—were the most productive and sustainable method for growing food. Fifty-two countries endorsed the findings.

"We've assumed for a long time that genetic diversity and organically rich soils are necessary," Herren told me. "But now we had the evidence as to why it actually works."

Then they hit a stone wall: the United States. The major seed-chemical companies and their allies in the Bush administration objected. The United States registered an official "dissent" claiming that the global team failed to recognize the contribution of liberalized markets and open-trade agreements in spurring innovation, and were overly skeptical of GMOs as a tool for addressing hunger. The United States and its allies Australia and Canada, both of which had conservative governments at the time, pressured other countries to withdraw from the findings.

"They were afraid that if they signed off on it, they'd be obliged to actually implement our recommendations," Herren said, still bristling eight years later at the recollection. "There was a fear that somebody might look at our recommendations and say, 'okay what are you doing now?' They'd be obliged to act." In the midst of the careening economic crisis in 2008 and 2009, the report was buried.

In the food universe, though, the consensus reached by a broad spectrum of scientists sent a message that it was time to find a new way forward for agriculture. Its effect was akin to the early climate reports from the United Nations's panel of international scientists, which helped trigger awareness about the depredations of climate change and the need for alternatives to fossil fuels. "This was our IPCC report for the food movement," commented Anna Lappé, a food policy activist and author of the path-breaking book, *Diet for a Hot Planet*. It was a roadmap, she said, for reforming agriculture.

**SHORTLY AFTER MEETING** Hans Herren, I met Kiersten Stead, a vice president in Monsanto's venture capital division, Monsanto Ventures. I asked her how the company's seed breeders are responding to the volatile growing conditions. She said that the company has been "breeding for climate change since the early two-thousands." In company test plots and hothouses, the company is prioritizing new genetically engineered traits to resist the new pests and diseases that are following the changing temperatures. She also identified a new generation of seeds being engineered to absorb more nitrogen from the soil to reduce their need for nitrogen-based synthetic fertilizers. Excess nitrogen fertilizer runoff from factory farms in the Midwest have been causing massive contamination of drinking water supplies; Stead says that the company is anticipating tighter regulations governing their use. Simultaneously, she said, the company is responding to rising reports of resistance to their own chemical products.

"We started looking into multiple traits that could be delivered at the same time," she told me. "We wanted multiple modes of action so if something was resistant, we'd have another trait stacked in—like an oncology cocktail, with multiple modes of action." That's called 'stacking'—which means 'stacking' multiple genetically engineered traits into a single seed—for example, resistance to the toxic effects of glyphosate in addition to resistance to the new generation of herbicides like Dicamba. And when it comes to insect pests, the company has been, "addressing above ground insect pressure and below ground insect pressure."[4] Meaning, engineering different toxins to fend off different insect pests that may attack different parts of the plant, and potentially stacking' those with genes to convey resistance to glyphosate applications and Dicamba applications—from two to five traits "stacked" in the same engineered seed.

In sum, she told me, "We work to uncouple the farm from the environment around it."

Let's call it a conscious un-coupling of the seed from the environment of the farm. This, I thought, was a pretty succinct description of the industrial agriculture paradigm.

**LONG BEFORE MONSANTO** became a household name, starting in the 1960s we consumers were uncoupling ourselves from the sources of our food as we moved *en masse* from rural areas into cities. Just as farms were being un-coupled from their surrounding ecosystems, we were uncoupling from our farms. We entered the age of synthetic chemical pesticides and fertilizers with our eyes wide shut.

And it worked, for a while: yields increased significantly over the decades between 1960 and the end of the century. Food became cheaper and far more abundant.

But we've been dealing with the collateral impacts from this approach for over half a century. It was thanks to the pioneering work of the great scientist-journalist Rachel Carson, and her many successors, that we have come to understand just how damaging the bargain has been—to the ecosystems that support our ability to conjure food from the earth, and to the health of those who do that work. Carson herself was a radical in a careful, deeply scientific way. She spent twenty years working as a marine biologist for the US Fish & Wildlife Service, and published two books in the 1950s describing the mysteries of marine life and the threats to our oceans. Then, in 1962, she applied her scientific skills above ground in her classic investigative book *Silent Spring*, revealing how pesticides like DDT and other neuro-toxins were severely damaging the health of human and non-human animals alike. (It would emerge later how damaging our reliance on chemical agriculture has been to the atmosphere).

I reread *Silent Spring* and was struck at how far ahead of her time Carson was in uncoupling the great promises of chemical advances in agriculture, celebrated in agricultural journals and by the press, from the realities on the ground. She revealed the dark side of the miracle chemicals, which were extremely effective in killing pests—many of them were initially devised as nerve gases by the Department of Defense—but had devastating impacts on the broader eco-system.

Since then, new generations of scientists have begun to question not only the health implications of chemical-based agriculture, but whether it's the most effective strategy for ensuring a reliable food supply. Scientists, often in spite of themselves, are becoming key players in the rapidly expanding movement to challenge the industrial paradigm that treats the seed as a foreign entity to be inserted into a chemically reconstituted environment. "The evidence," Herren said, "is prying open cracks in the 'cathedral.'"

From the flat expanses of the American Southwest to the Midwest to the rolling hills of west Pennsylvania, from the terraces of southern France to the majestic Pamir Mountains of Tajikistan, to Ethiopia and Tunisia and the drought depleted factory farms of California's Central Valley, and in many other diverse locations across a spectrum of ecological conditions, the results from our experiment with industrial agriculture have been rolling in. The signs of environmental stress are rising and the responses to them are taking shape. It turns out that separating the seed from the conditions in the field may not be the most reliable way to grow food. Farms which rely on seeds that respond to ecological forces, rather than attempt to overwhelm them, can not only produce abundant quantities of food, but have far greater resilience in the face of dramatic shifts in growing conditions.

Since Herren's ground-breaking report, dozens of peer-reviewed scientific studies have highlighted the importance of a diverse source-pool of seeds planted in soils rich with organic matter to ensure resilience to ecological disruption. Even the FAO, the "cathedral" itself, has been moving toward adopting some of the basic principles of agricultural ecology in its international development programs. "By introducing agroecology principles, you can reduce the risks of exposure to climate change," a top FAO official declared during an international agroecology symposium the organization sponsored in

April 2018.[5] Of equal significance is the mounting evidence that biologically rich soils, saturated not in chemicals but in organic material, can reverse agriculture's role as a contributor of greenhouse gases to sinkholes for absorbing $CO_2$ from the atmosphere, providing more carbon-rich soils in the process.[6]

Fifty-year old practices once considered "modern" are looking more and more old-fashioned as their vulnerabilities become clear. "There isn't anything particularly innovative about chemical agriculture," said Anna Lappé.[7] "They have nothing to do with innovation. They're a set of practices that are particularly profit enhancing because it involves proprietary chemicals, many of which have been around for decades, and proprietary seeds that require farmers to buy them year after year."

Soil, it turns out, is not just dirt waiting to be tamed into submission for crops. Farm practices once considered to be on the margins of the big-ag economy—once dismissed as "hippie farmers"—are turning out to be highly effective for dealing with extreme levels of climate stress. Healthy soil teems with billions of micro-organisms and minerals which are beneficial to plant life.[8] A long-term comparison by researchers at UC Berkeley of soil in fields with varying levels of organic materials revealed a direct correlation between the population of soil micro-organisms and increases in soil fertility, the ability to resist attacks of pests and diseases, and enhanced capacity for water absorption—all of critical importance as rainfall becomes more sporadic.[9] Claire Kremen, chief author of the study and head of the Berkeley Food Institute at UC Berkeley, which is pioneering new ecologically regenerative growing practices aimed at long-term food sustainability, commented: "The inter-relationships in more diverse and organically rich fields mean they're much more able to defend themselves against new threats and changing conditions." In many instances, they're also able to produce comparable quantities of food.

The most comprehensive side by side comparison of organic vs. conventional agriculture occurred over thirty years in the rolling hills outside the town of Kutztown in southeast Pennsylvania. From 1981 to 2011, researchers from the Rodale Institute planted fields using three different cropping methods: organic with composted manure for fertilizer; organic compost made from other plants; and conventional, utilizing common chemical pesticides, herbicides, and nitrogen-based fertilizers. They planted corn, soybeans, and wheat, all common to the area, as well as cover crops in the organic fields like rye and other grains. The results: yields on organic plots, averaged over three decades, either equaled or, in some instances, surpassed those in conventional plots. This was the case even though the organic had "higher levels of weed competition" than their conventional counterparts—which, they surmised, was because crops growing in soil that was richer with organic material could "tolerate much higher levels of weed competition." Another finding: that the yields in organic fields were not just equal but they were higher in four out of the five "drought years" experienced over the course of the trial— which they attributed to "higher water holding capacity of organic soils."[10] The major downside on organic was that farmers had to do about 25 percent more labor, reflecting extra time spent in the field monitoring pests, removing weeds, preparing the soil, and planting cover crops—activities which in conventional fields are accomplished with chemical fertilizers and pesticides. Overall, this study is seen as a pathbreaker, demonstrating the potential for organic to deliver substantive yields over time while accounting for the normal ebbs and flows of weather and market conditions.

Comparative yields can also vary widely depending on the crop. Other studies show that while the production of many fruits, tomatoes, and legumes like beans, peas, and peanuts can come close to par with conventional yields, cereals like oats and some vegetables can have as much as one-third less yield per acre, as can wheat grown in less-attentive circumstances than the Rodale farm.[11] The distinctions arise from numerous factors—the amounts of nitrogen in the soil, the presence of weeds and pests, and the type of seed used. Such side by side comparisons have been rare until recently, and may offer a skewed picture due to the enormous discrepancies in research funding. Over the past three decades, Claire Kremen pointed out, organic agriculture has received barely 2 percent of the USDA's overall research budget. If that were even closely on par with conventional R&D, she said, organic could get a lot closer to parity across a range of crops. And it's already pretty close in some crops with techniques developed over many seasons by farmers themselves with little help from the government.

Seed diversity can make a big difference in controlling pests without pesticides. Researchers from South Dakota State University found a direct link between the range of biodiversity of plant and animal populations and "significantly reduced pest populations" on a cross-section of corn farms in the heart of Midwestern mono-crop farm country. The higher the diversity, the lower the pest population. This counter-intuitive conclusion, they surmised, was because in a field less reliant on chemicals, predators can help keep crop pests in check, and there is more food available for a diverse array of creatures so that pest populations do not dominate. The reverse, they found, is also true: less diversity means fewer pest predators and more uniform crops on which pests can feast.[12]

Hurricane after hurricane, starting with Hurricane Mitch that blitzed through Central America in 1998, have provided insights into which farms survive in extreme conditions. Miguel Altieri, who for many years ran the agricultural ecology program at UC Berkeley, conducted a multi-year study of resilience to hurricanes like Mitch and found that the farms which recover quickest from such devastating events were farms with more diverse fields, while industrial-scale monocultures frequently lost everything and took multiple seasons to resume full production. The former showed significantly less loss of topsoil, greater ability to absorb water, and far less erosion than nearby conventional, mono-crop farms.[13] The latter turned out to be highly fragile when conditions veered from the "norm"—a norm which our distressed atmosphere is no longer delivering.

In the American Midwest, according to a team of researchers at Kansas State University, wheat farms are estimated to face a drop in production of as much as 6 percent for each degree Celsius the temperature rises.[14] A key mitigating factor to face the "escalating challenges presented by climate change," they concluded, is a greater diversity of seeds "to ensure future food security."[15] Conversely, a major consequence of "increased crop homogeneity is the potential for yield instability." Erratic yields are more likely, they conclude, with the "increased unpredictability in weather patterns associated with climate change." On farms with a more diverse variety of seeds, some plants will die but others will survive, and they become the progenitors of new generations which contain the hardier genes.

Stresses on the industrial food system are heightening pressure on the publicly supported crop insurance system in the United States. Between 2011 and 2016, crop insurance payouts jumped from $10 billion to more than $17 billion; at least 60 percent of those costs were borne by taxpayers. Most of the indemnities were due to factors linked by the (Obama-era) USDA to climate change—namely heat, drought, and, on the flipside, too much rain at the wrong time.[16] The highest rates of increase were in the Midwest and California's Central Valley, where endless miles of identical crops have been unable to adapt. The system is perilously close to collapse due to liabilities linked to climate change, according to the Government Accountability Office (GAO), the congressional auditor of US federal programs, which has identified crop insurance as high on the list of federal programs at high risk of experiencing major financial crises.[17]

The situation is likely even worse than the GAO's ominous forecast due to the curious rules of crop insurance that enable many farmers to shield their losses from public view. In America's wheat belt—Kansas, Oklahoma, Nebraska, and Texas—the USDA has tried to make it easier for farmers to obtain crop insurance by allowing them to write-off as many as seven of their worst production years in their applications for revenue insurance, which is designed to cover shortfalls. Such seven-year exemptions in calculating an average income, serves to mask the performance of seeds that perform badly, explained Anne Weir-Schechinger, the co-author of an investigation into the practices by the Environmental Working Group. "If you can exclude bad years, it means we're not seeing when there are crop losses." The result, she said, is repeated planting of failed seeds that could lead to another Dust Bowl in the Midwest as depleted soils swirl in the intensifying dry winds.[18]

The big jump in crop-insurance payouts was a wake-up call at the Obama-era USDA. Long the stronghold of big-ag boosters, the agency for the first time started to recognize the need for new approaches to farming as climate change ups the financial consequences. Funding for organic agriculture was quadrupled, and new recommendations for farmers were disseminated that sounded like they were ripped from a primer for one of those "hippie farms." Declared the USDA: "[D]iversifying crop rotations, integrating livestock with crop production systems, improving soil quality, minimizing off-farm flow of nutrients and pesticides, and other practices typically associated with sustainable agriculture are actions that may increase the capacity of the agricultural system to minimize the effects of climate change on productivity."[19] In other words, the way that industrial farming is conducted today is precisely the wrong approach for cultivating crops that are resilient to climatic disruptions.

Under President Trump, however, it's unlikely we'll hear the two words "climate" and "change," or for that matter "disruption," coming from the USDA, not to mention any other federal agency in a government now headed by a climate denier. Just when we need more independent science probing into how best to grow our food, the Trump administration is steering us headlong toward a false reality rather than the one we are experiencing.

But the cat's out of the bag. The official denials may mount, but reality keeps intruding. If the government doesn't recognize the profound climatic changes, farmers certainly do; they're contending with them daily. Sometimes they may not even call it "climate change" but, as one Central Valley farmer told me, "Something out there is changing."

Hans Herren comments: "We're talking about changing the paradigm. Old science was the Green Revolution. The new science is agro-ecology. The way to deal with climate change is more diversity, not less."

## SEED POPULATION EXPLOSION

In June 2010, Salvatore Ceccarelli, an Italian agronomist, shipped two twenty-kilogram sacks of wheat seeds from the seed bank outside of Aleppo onto a plane headed back to Italy. Seven years later, the seeds he spirited out of Syria as the country descended into a brutal civil war took root on Italian farms—and have gone on to present a potential challenge to the chemical companies' dominion over the world's seed patents.

Ceccarelli is one of the world's most renowned seed scholars and practitioners. He's done TED talks in Italy, consulted with governments around the world on policies to encourage biodiversity, and has farmed and researched seeds in hot dry places as far afield as Ethiopia, India, the Jordan Valley, Italy, and California. He was based for most of the past quarter-century at the ICARDA seed center in Tal Hadya, outside of Aleppo, as director of the breeding programs for wheat and barley—the very same place that supplied those Hessian fly resistant seeds to farmers and researchers in Kansas. Among Ceccarelli's many duties over the years, he trained young plant scientists in Abu Ghraib before the seed bank was destroyed during the US invasion. He's an outspoken advocate of what he calls "participatory plant breeding"—engaging farmers themselves in breeding new crop varieties, rather than leaving that to the consolidating group of global seed companies. Ceccarelli was gone from Syria by the time the last scientists were forced to evacuate the seed bank to Lebanon, but he had ensured that a part of its legacy lives on in Italy.

I caught up with Ceccarelli the day after he'd touched down in San Francisco and was en route to deliver a keynote address at a seed conference at UC Davis. He's tall and wiry, and told me enthusiastically about how those two sacks of Syrian seeds are poised to upend the global plant patent system.[20]

Inside each of the two sacks of seed that Ceccarelli sent to Italy were dozens of different wheat varieties—not single seeds, but a mixture of seed populations, like they'd been planting and harvesting for centuries in Syria. In Italy, he passed the seeds along to the Rural Seed Network (Reto Semali Rurali, RSR), one of Italy's oldest agricultural reform NGOs. The RSR passed the seeds along to two of their member farmers—one bag went to a farmer in Sicily, Giuseppe li Rosi, and the other to a farmer in Tuscany, Rosario Floriddia.

Meanwhile, in Brussels, the RSR and a coalition of environmental NGOs were lobbying the European Council—the European Union's executive body—to change a key European law governing seeds. The law, like a similar one in the United States, requires commercially available seeds to be certified as uniform and stable with clearly identifiable characteristics. In other words, certified that they do not change. This provision has been critical to facilitating the entry of major corporations into seeds, for only a seed that remains unvaryingly distinct in the same way year after year can be patented, and thus branded. But this rigidity, the coalition argued, makes for seeds that are uniquely unsuitable to volatile growing conditions—a hot-button matter in Italy and throughout much of Europe, which has been facing record-breaking temperatures and, in some regions including Italy, a multi-year drought.

In 2015, they succeeded. The European Union agreed to waive the registration requirements on four crops, and to launch what it described as "a temporary experiment . . . for the marketing of populations of the plant species wheat, barley, maize and oats." Mixtures of seeds, evolving and changing and sharing genes in the ways that plants do naturally, could be officially registered for sale. Enter the Syrian seeds. "These populations are extremely dynamic, they are responding to natural processes happening all of the time," Ceccarelli told me.

Within four growing seasons, the two populations of Syrian seeds grown in different parts of Italy showed significantly different characteristics, a live lesson in "Evolution 101." In Sicily, with a fraction of the rainfall of Tuscany, the wheat matures several weeks earlier and is as many as two to four inches shorter than the Tuscan varieties. Those varieties, in a more moderate and moist climate, mature later and deliver more protein per plant. "Compare the two in the same environment," Ceccarelli told me, "and it's day and night." He argues it's the diversity that gives fields the ability to adapt to new conditions. "Explain to me how a crop that is uniform and stable responds to climate change . . . Today if you are a dynamic seed company you are working on varieties for 2025. For which sort of climate? How many more degrees hotter will it be? Do they know what pests and diseases will come with the new conditions? These population mixtures are the cheapest and most dynamic way to cope with climate change."

The Italian agriculture ministry sent an inspector to the Tuscan and Sicilian fields in June 2017. She made a historic determination: the seeds passed a rigorous set of rules to ensure that they were of consistent quality and free of disease, and certified them both. For the first time since the rules were written more than a quarter-century ago, a mixture of seeds was approved for sale in Europe. Distinctions were made based on geographic reality: Mr. De Ruti was authorized to sell up to two tons per year of the seeds cultivated in Sicily, and Mr. Floriddia to sell up to three tons per year of the seeds cultivated in Tuscany—reflecting the different yields of each of the two distinct populations, which had the same origin but were growing in very different locales.

Riccardo Francolini, a staff member of the RSR, recalled by phone from the group's headquarters outside Florence the drama of witnessing a seed inspector doing what they'd never done before: "They're used to seeing a single variety, all the same in a field. It was something amazing for them to stand in front of something so different. The idea of a 'population' changes the vision in a profound way."

That shift requires taking a different view of seeds in a field: Are they delivering one or perhaps two desired characteristics, like the conventionally bred varieties? Or are they delivering the capacity to express a range of characteristics, some of which may not be immediately apparent, like the jumble of seeds in the Italian fields? Those, Ceccarelli and his allies assert, may include traits critical to surviving climatic extremes.

Within a year, at least one hundred farmers were growing the Syrian wheat seeds organically in Italy, according to Ceccarelli. The yields don't match the yields of neighboring farms, which require intensive synthetic chemical inputs. But what they show, he said, is of equal or greater importance: "High rates of yield stability, year in and year out." As the temperatures went up and the rainfalls went down— and then up again—they kept producing, without the assistance of expensive and corrosive synthetic boosters. The bread and pastas made with the wheat have a budding market, including among those with gluten intolerance, which these varieties do not appear to trigger.

More seed populations from other parent lines were also approved across Europe. At least twenty of what are technically called "cross-composite populations" were authorized by national seed authorities in the United Kingdom, Germany, Denmark, and France, representing a total of about four hundred tons of seed. Klaus Rapf, a board member and adviser to *Arche Noah* (Noah's Ark), a seed-saving and research institution in Austria that's been part of the coalition fighting for the reforms, considers this a momentous shift. "What's at stake is the very concept of 'variety,'" he told me. "Defining something as a 'variety,' unchanging, is a concept created to defend turning a seed into a protected intellectual property, based on the notion of very high uniformity." The seed populations, he said, challenge the reigning presumption that we can breed our way out of the crisis in our food-growing lands with lightning strikes of engineered or hybridized seeds, bred to specific conditions which are dynamic and likely to rapidly change. By contrast, it's the breadth and diversity embodied in those populations that hold the key for adaptability to varying and often adverse conditions.

"[Plant] patents," Ceccarelli said, "are a way for the seed industry to control evolution. The seed companies only make money by selling one variety to massive amounts of farmers. Their only possibility of evolving is to replace a variety one year with another one the next year. You can't adapt to climate change that way." And you can't, he said, patent a population that is in constant interplay with the environment. "We are registering and certifying something that is evolving. Next year will be different, and the year after that. You start with one thing and you end up with another thing totally different."

The experiment with seed populations was given a three-year lifespan; at the end of 2018, the European Union will assess the results and whether it will continue or expand. The stakes are high. If renewed, the global nature of the seed business—the three dominant companies that sell to Europe and the United States and across the globe—ensures that it would not take long for the principles underlying composite seed populations to make their way into the United States, where there are already independent field trials of "evolutionary populations" underway. Those fields in Europe, started with a couple of bags of Syrian wheat seeds, could be a slow-burning fuse on global plant patents.

# CHAPTER 8

# SEEDS: THE ELEPHANT AND THE ACORN

**TALK ABOUT SEEDS** long enough and you'll meet the elephant in the room—the specter of hunger. Can we produce enough food for the earth's growing population? As the Biotechnology Innovation Organization, the bio tech industry's trade group, puts it: "To feed everyone, we'll need to double the amount of food we currently produce."[1] That formulation is pretty much a mantra for every one of the group's agri-business company members. A "starving African," somewhere, has in various forms been the meme for explaining why the end goal of all agriculture is maximizing yield no matter the consequences.

But shaping the challenge like a Malthusian equation—more food for more people—is also a dodge around the far more fundamental question: Who eats the food that is produced? For whom do the seeds of the world toil?

The quantity of food grown from the world's seeds has little bearing on who eats it. From New York hedge funds to Chinese banks to sovereign wealth funds of the Middle East, big money is moving not only into seeds, but into agricultural land. Millions of fertile acres in Africa and Asia are being diverted into export plantations to feed the people in the richest countries.[2] Large populations of the officially "under-nourished" in Africa, Asia, and Latin America live within striking distance of large farms devoted to exporting vegetables and fruits to the United States, Europe, the Persian Gulf states, and China.

North Americans need not go far to see the inequities of food distribution in action. One of the most food insecure areas in the United States is the country's biggest center of fruit and vegetable production: California's Central Valley. Amidst abundant fields of fruit and vegetables, low-wage farmworkers are frequently unable to afford the food that they pick, and one in three children in some of the agricultural towns there go to bed hungry.[3] Across the United States, countless billions of corn seeds—40 percent of them—give birth not to food for humans, but food for automobiles, in the form of ethanol.[4] Another third of corn and other grains, it pains me to say as a devoted omnivore, is devoted to feeding livestock, not people.[5] More are turned into processed foods—chips and soft drinks and sugared cereals and so on—which offer little nutritional value and contribute to rising rates of obesity. Thanks to the wide discrepancies between seeds planted and food consumed, more than fifteen million US households, in one of the wealthiest countries on earth, are considered "food insecure."[6]

In developing countries, the primary reason for 795 million people being officially "under-nourished," according to the FAO, is not the inadequate quantity of food being produced. Rather, it's what the FAO calls "less inclusive economic growth."[7] That's diplo-speak for the unequal distribution of resources—in this instance, the most essential of all resources, food. Hungry people are often surrounded by food—on farms, in shops, and in open markets—but can't access it.

The United Nations special rapporteur on the right to food, deputized by the UN General Assembly to fight hunger, vigorously disputes the assertion that the seed-chemical nexus is necessary to feed the growing global population.[8] The organization responsible for ensuring that millions of hungry people actually do have enough to eat—and which has no financial incentive to promote one agricultural model over another—says that there's plenty of food produced to feed several billion more people than our current global population of seven billion. In early 2017, the rapporteur released a report accusing the companies that manufacture pesticides—the very same companies that manufacture seeds—of "systematic denial of harms," of "aggressive, unethical marketing practices," and obstructing efforts by governments to apply more substantive oversight of their use.[9] Meanwhile, the devastating health toll taken by agri-chemicals continues to mount—two hundred thousand deaths a year from acute chemical poisoning, according to the World Health Organization, and that's just talking about humans, and not the complex web of creatures and micro-organisms that live in and around farms. Such systemic chemical poisonings are most directly suffered by those who are doing the work to provide food to others.

"The 'food problem' is not a production problem," says Eric Holt-Gimenez, executive director of Food First, an NGO that probes into the underlying causes of hunger. "It's a problem of income. There's not a 'food' problem, there's a poverty problem."

The journalist Michael Pollan, whose writing and reporting helped lay the groundwork for the global food movement, says that financial props actually encourage farmers to over-produce, which help commodity companies keep prices low. "Everybody's convinced that agriculture is a scarcity economy," he told me. "That's a big myth. The question is, who eats what?"

So, the question of which seed to plant—or for most of us, far more precisely, which seeds others plant on our behalf—is not a matter of deciding between which food-stressed person we hope to help feed. Deciding which seeds to plant is a decision about what kind of agriculture will be used to grow and cultivate them into food. As Talavai Denipah-Cook, a member of the Hopi tribe who handles ecological matters on the Ohkay Owingeh Pueblo in New Mexico, told me: "The seed is not just the seed. It's how you plant, it's how you treat the soil you plant it in. And once you start, it becomes how do you live on the earth."

The one-eyed focus on yield to the exclusion of other factors has led to practices which degrade the conditions on which the long-term vitality of our agriculture depends. For decades, the deck has been stacked. Industrial agriculture interests have benefitted from billions of dollars in public funds for research and development, and billions more for insurance policies in case the strategies don't work. There are, however, other practices, which begin with more organically rich soils and seeds that interact with the ecological homes where they're planted, which have been shown to be far less destructive to the earth's ecological and atmospheric balancing act, and far more resilient to the tumultuous changes underway in our food growing lands.

Those who argue that we must jump on the chemical-seed treadmill in order to "feed the hungry" are using it as a moral fig-leaf, diverting us from the fundamental matter of *distribution* of the food we already produce. Food security is not an issue of quantity, but of location—meaning one's location on the geographic map and on the socio-economic ladder.

The chemical-seed combines have spent the last thirty years squeezing thousands of seed varieties off of farms, and out of the currents of evolutionary history. Many of the seeds displaced from their natural ecological locales are now, ironically enough, being housed in the mountain freezer in Svalbard. Those seeds are frozen in time—a time that may never be repeated.

**THE MOUNTAIN OF** Svalbard, located a one-hour plane ride north of the northernmost Norwegian city of Tromso, was not supposed to melt. But it is the improbable that is becoming more likely every day. The expanding spectrum of unknowns is such a defining feature of our times that scientists have characterized it with an unnerving three words—"Stationarity is dead."[10] That was the title of a groundbreaking paper in *Science* by a multinational team of water scientists. "Stationarity" is the term scientists use to refer to the range of variation in natural conditions over time that serve as a kind of baseline, from which we can measure change. They studied the records of rain fall on the planet going back thousands of years—they went as far as data analyses in a field known as paleo-hydrology can go. They concluded that the patterns we're seeing today bears little or no relation to what we've seen in the past. Greenhouse gases are wreaking havoc on planetary rhythms. Predicting the future based on the past or the present is no longer possible, Christopher Milly, chief author of the paper and a senior analyst with the US Geological Survey in Bethesda, Maryland, told me. "[C]limate change comes along," he said, "and the textbook examples are no longer relevant to what we've been observing in nature."

A frantic search is underway for seeds that can keep up with these onrushing changes. As I write, there is no doubt another buzzword being conjured in a marketing room somewhere for seeds that can be promoted as bred for the volatile contingencies of climate change. The current slogan is "Climate Smart Agriculture." Needless to say, any strategy that does not recognize how profoundly climate change is altering conditions for agriculture would definitely not qualify as "smart." But the very concept demands questions: "smart" for whom? "Smart" for how long? Are chemicals necessary to enable them to grow? Who owns those? And, perhaps most importantly, are there alternatives that are less destructive to the earth's fragile ecology and the public's health that will deliver similar characteristics without the collateral damage?

The new frontier of high-tech seed research and development moves beyond inserting genes from other species, as with GMOs, but to genetic manipulation within the genome itself—the technology known as CRISPR. It involves adding gene drivers to activate one or another characteristic, and is supposed to be simpler and less genetically disruptive than GMOs. The principles, though, are similar—manipulation of the genome to obtain a desired trait. And similar concerns have been raised with both forms of gene manipulation—high unknowns as to the environmental or health implications, and definite knowns that the conditions for which they're being bred will change. Whatever the next iteration of "climate smart" agriculture, another thing is certain: the more distanced we permit ourselves to be from the food growing process, the more we will be dependent on Bayer-Monsanto, DowDuPont, and Syngenta-ChemChina for our food.

There is no longer a steady natural state for which seeds can be bred. What happens if a drought gives way to abundant rainfall—as occurred in California in 2016 after a record-breaking dry spell of five years? Or, what if you've bred or genetically invented resistance to one pest, and another, driven by the heat, shows up? Or you store seeds in a frozen mountain that starts to melt? The volatility represented by "the end of stationarity"[11] means these questions must be asked on the ground, farm by farm, seed by seed.

"'Climate-friendly crops' is a misnomer," said Gary Nabhan, the desert agriculture specialist in southern Arizona. "How can you be 'friendly' to uncertainty? How is a plant breeder to know how much heat or drought resistance to breed in, or what kind of resistance to which pests and diseases...? Seed diversity must be integrated with more diversified farm systems. There is no silver bullet. You have to breed for uncertainty."

Uncertainty is the worst possible set of conditions for a tightly controlled industry which requires predictability on a mass scale— precisely what climate change is turning into a relic of the past.

Svalbard's seeds, and the seeds in Fort Collins and at similar global institutions, poised on their ice-cold steel shelves, are in many ways a stand-in for the seeds of the world—making a last stand for diversity threatened by climate chaos in the clouds, corporate consolidation on the ground, and patent laws that remove them from the slipstreams of evolution. Svalbard's vulnerability makes it even more imperative that living seeds are engaged in a dynamic interchange with the rapidly changing environment and not just backed up in a frozen vault.

"The best way for preserving this genetic material is to grow these seed populations out so that they're evolving all over the world," said Salvatore Ceccarelli, who is doing that with the seeds he brought from Syria to Italy. That's why Ahmed Amri is tending to the fields of displaced Fertile Crescent seeds in Lebanon and Morocco; why Robin Pailzote of the White Mountain Apache is tending to the descendants of seeds planted by her ancestors in southern Arizona; why Justine Hernandez, like other seed librarians, is collecting and cataloging seeds at the library in Tucson; why Bill McDorman is seeking out and planting local cereals and grains in the western Rocky Mountains; why the Navdanya seed network is flourishing with more than fifty branches throughout India; why the Seed Savers Exchange in Decorah, Iowa, has expanded from a small hobbyhorse into a force for seed diversity in the dead-center of Midwestern industrial-agriculture. It's why the official seed centers like ICARDA and CIMMYT in Mexico are moving as quickly as they can to preserve threatened seeds; why dozens of new local and organic seed companies have defied the trends of consolidation since 2010; and why countless millions of farmers are cultivating, tending, exchanging, and breeding out their seeds. They are all central to sustaining the vitality of our food resources and contributing to all of our food security. And they are all contending with the zig-zagging rhythms made dissonant and jerky and unreliable by the suffusion of greenhouse gases into our atmosphere.

There's a twist, too. All seeds on our planet are being subjected to the same pressures—including the multifarious indigenous seeds cultivated by all these extraordinary people and institutions. As Joy Hought, executive director of Native Seeds/SEARCH, told me: "The thing with climate change, it's not so much what we know so far. It's what we don't know at all. Some of these older varieties, just like the conventional crops, they've also never been subjected to the stress at this level. The crops we grow, and the ones they grow on the big industrial farms—none have been subjected to the stresses of climate change that we expect are coming."

All of us are entering uncharted territory. There's a lot of adapting we all need to do across a number of fronts. When it comes to food, the most essential of all resources, the stories contained in our seeds are posing the same question to all of us: Should we hitch our food security, and the food security of future generations, to seeds bred and marketed by three multinational chemical companies?

**FOR THE PLANET,** the uncertainties we're experiencing amount to the symptoms of ecological trauma. That trauma is unfolding in slow but accelerating motion on what had been a pretty well-balanced planet, at least until greenhouse gases started concentrating in the atmosphere. Now the balance is deeply disturbed, the largely invisible push-pull ebb and flow of gases in the atmosphere above our heads is in turmoil. The feedback loop from the unpredictable churn above our heads is rattling the connective circuits that link all living organisms to one another.

When it comes to responding to trauma, the principles of "resilience" in seeds and ecosystems are not that different from the resilience we humans need to contend with sudden changes in our circumstance, sense of place, and loss of loved ones. I've experienced my share (like many of us, I imagine). Understanding how one recovers from trauma as a human being is not that different from trying to understand how ecological systems respond to trauma. They, like humans, need diverse networks of support. In the case of seeds, this means plenty of water and nutrients in the soil, crops that return those nutrients to the soil after they've been depleted, and plenty of diverse options to up the chances of survival in changed circumstances.

My favorite description of seeds comes from the great Irish playwright George Bernard Shaw, who characterized them with just three words: seeds contain, he said, "a fierce energy."[12] Shaw was talking specifically about acorns and the marvel of how they grow from tiny kernels into mighty trees dispensing tasty nuggets, but the same words could apply to any seeds on earth.

Seeds drop from trees, blow from flowers, tumble from the feet of bees, arise from the waste of squirrels. A kernel can spend centuries lying dormant, apparently lifeless (like that two thousand-year-old date palm harkening back to the Roman empire). Then add some water, a dapple of sunshine, a couple of minerals, and voilà: a stream of energy commences. Seeds actually sense the presence of those life-giving elements and that it's time to start photo-synthesizing, time to send sugars through its veins, time to emerge. The shell cracks, a sprig emerges, a stem pokes above the surface of the earth and delivers to us a flower or a fruit or more seeds that provide food for humans and other animals.

It takes a certain ferocity to accomplish this transformation from dormancy to life —as it does to adapt to the changes that are occurring in the ground in which seeds grow. The metabolism of seeds growing in the midst of drought will slow down to preserve their energy until conditions change; leaves contract or unfurl depending on the flow of water and intensity of the sun, for just two examples among many. Seeds can control those basic adaptive functions. But the question now is: Who controls the seeds?

They are where the story ahead begins.

# BIBLIOGRAPHY

Here are some books which, to one degree or another, inform the reporting and the thinking in these pages. For those interested in pursuing any of these themes more deeply, all provide new and fruitful insights.

## ON NIKOLAI VAVILOV AND LUTHER BURBANK

Dreyer, Peter, *A Gardener Touched by Genius: The Life of Luther Burbank* (Luther Burbank Home & Gardens, New and Expanded Edition, 2002).

Nabhan, Gary Paul, *Where Our Food Comes From: Retracing Nikolay Vavilov's Quest to End Famine* (Washington, DC: Island Press, 2009).

Pringle, Peter, *The Murder of Nikolai Vavilov: The Story of Stalin's Persecution of One of the Great Scientists of the Twentieth Century* (New York, NY: Simon & Schuster, 2008).

Smith, Jane S., *The Garden of Invention: Luther Burbank and the Business of Breeding Plants,* (Penguin Press, 2009).

## ON FOOD AND A SUSTAINABLE FOOD SYSTEM

Allen, Arthur, *Ripe: The Search for the Perfect Tomato* (Counterpoint, 2011).

Lappé, Anna, *Diet for A Hot Planet: The Climate Crisis at the End of Your Fork and What You Can Do About It*, (Bloomsbury USA, 2011).

Pollan, Michael, *Food Rules: An Eaters Manual*, illustrations by Maira Kalman, illustrator (Penguin Press, 2011).

## ON THE SEED-CHEMICAL COMPANIES

Gilliam, Carey, *Whitewash: The Story of a Weed Killer, Cancer, and the Corruption of Science* (Washington, DC: Island Press, 2017).

Kloppenberg, Jr., Jack, *First the Seed: The Political Economy of Plant Biotechnology* (University of Wisconsin Press, 2004).

Krimsky, Sheldon, and Jeremy Gruber, eds., *The GMO Deception: What You Need to Know about the Food, Corporations, and Government Agencies Putting Our Families and Our Environment at Risk* (New York, NY: Skyhorse Publishing, 2014).

## ON THE BIOLOGY, ECOLOGY, AND PRESERVATION OF SEEDS

Chamowitz, Daniel, *What A Plant Knows: A Field Guide to the Senses*, Updated and Revised edition (Scientific American/Farrar, Straus & Giroux, 2017). (Collects in one book much of the current knowledge about the responsiveness of plants to their environment.)

Fowler, Cary, *Seeds on Ice: Svalbard and the Global Seed Vault* (Prospecta Press, 2016).

Hanson, Thor, *The Triumph of Seeds: How Grains, Nuts, Kernels, Pulses, & Pips Conquered the Plant Kingdom and Shaped Human History* (Basic Books, 2015).

Pollan, Michael, *The Botany of Desire: A Plant's Eye View of the World* (New York, NY: Random House, 2002).

# BIBLIOGRAPHY

## ON THE FUTURE OF FOOD AND AGRICULTURAL STRATEGIES

Allen, Will, *The War on Bugs* (Chelsea Green, 2008).

Bourne, Joel K., *The End of Plenty: The Race to Feed a Crowded World* (WW. Norton & Co, 2015).

Brescia, Steve, ed., *Fertile Ground: Scaling Agroecology from the Ground Up* (Food First/Institute for Food & Development Policy, 2017).

Imhoff, with Foreword by Michael Pollan, *Food Fight: The Citizen's Guide to the Next Farm Bill* (Watershed Media, 2012).

Orion, Tao, *Beyond the War on Invasive Species: A Permaculture Approach to Ecosystem Restoration* (Chelsea Green, 2015).

Williams, Justine M., and Eric Holt-Jimenez, eds., *Land Justice: Re-Imagining Land, Food, and the Commons in the United States* (Food First Books/Institute for Food and Development Policy, 2017).

## ON ECOLOGICAL PHILOSOPHY, SCIENCE, AND SEEDS

Griffin, Susan, *Woman and Nature: The Roaring Inside Her* (Sierra Club Books, 1978).

Kuhn, Thomas S., *The Structure of Scientific Revolutions*, 50th Anniversary Edition (University of Chicago Press, 2012).

Wall Kimmerer, Robin, *Braiding Sweetgrass: Indigenous Wisdom, Scientific Knowledge and the Teachings of Plants* (Milkweed Editions, 2015).

Williams, Terry Tempest, *When Women Were Birds: Fifty-Four Variations on Voice* (Picador, 2013).

# RESOURCES

When it comes to seeds, of course, everything discussed here and beyond is constantly evolving. I've included a list below of sources which focus in one way or another on seeds. This is not by any means an exclusive list, and inclusion in no way suggests an endorsement of their work.

## SEED SAVERS

Native Seeds/SEARCH (Tucson, AZ)
Southwest seed savers and research
*www.nativeseeds.org*

Seed Savers Exchange (Decorah, Iowa)
One of the largest seed saving and exchange institutions
*www.seedsavers.org*

Seed System
International NGO working to enhance seed security in vulnerable
   regions
*www.seedsystem.org*

## RESEARCH AND ADVOCACY FOR SEEDS

Center for Food Safety/Save Our Seeds initiative

NGO focused on agriculture reforms, and promoting seed saving

*https://www.centerforfoodsafety.org/issues/303/seeds/about-save
-our-seeds*


Public Eye

A Swiss NGO that tracks and investigates multinational companies, international aid programs and the actions of multilateral institutions involved with seeds.

*www.publiceye.ch*


U.S. Right to Know

Experts in document collection on the seed and chemical industries.

*www.usrtk.org*


Bioversity International

Supports research aimed at enhancing agricultural biodiversity

*www.bioversityinternational.org*


Union of Concerned Scientists, Food & Agriculture

NGO using science-based reports and monitoring of government food and environment programs

*https://www.ucsusa.org/food_and_agriculture#.Wt-EmEyZMpQ*


Indigenous Environmental Network

Coalition of native communities around the world active in preserving traditional seeds

*www.ienearth.org*

# RESOURCES

Food First

NGO and research center on underlying causes of hunger and
agro-ecological responses

*https://foodfirst.org/*

ETC Group

NGO monitoring the seed and agri-business companies

*www.etcgroup.org*

Institute for Agriculture and Trade Policy

A think tank probing into the impact of agriculture policies and trade
on food

*www.iatp.org*

Edible Schoolyard Project

Promotes gardens and the use of curriculum to teach the back-story to
our food

*https://edibleschoolyard.org*

## TRADE ASSOCIATIONS

American Seed Trade Association

Trade group representing the American seed industry

*www.betterseed.org*

European Seed Association

Trade group representing the European seed industry

*www.euroseeds.eu*

Organic Seed Alliance

Trade group and research center for producers of organic seeds.

*www.seedalliance.org*

Biotechnology Innovation Organization (BIO)

Trade group representing the bio-technology industry, including agricultural biotech.

*www.bio.org*

## NEWS & RESEARCH

Food & Water Watch

Conducts research into federal food programs and promotes reforms
of the agriculture and food system

*www.foodandwaterwatch.org*

Food Tank

A think tank focusing on agriculture and food issues

*https://foodtank.com*

Food and Environment Reporting Network (FERN)

Network of journalists producing in-depth reporting on food-
environment stories

*www.fern.org*

# RESOURCES

## AGRICULTURAL INSTITUTIONS

Food and Agriculture Organization of the United Nations
*www.fao.org*

United States Department of Agriculture
*www.usda.gov*

UNFCCC-Ag/climate
United Nations Framework Convention on Climate Change/
    Agriculture
*https://unfccc.int/topics/land-use/workstreams/agriculture*

Consultative Group for International Agricultural Research (CGIAR)
Oversees the UN's nine official seed banks and conducts research into
    new sustainable farming methods.
*www.cgiar.org*

African Center for Biological Diversity
Research and advocacy group promoting agroecology biodiversity in
    agriculture in Africa.
*https://www.acbio.org.za*

National Association of Plant Breeders
Professional and research organization for plant professional plant
    breeders
*www.plantbreeding.org*

# ACKNOWLEDGMENTS

I would like to acknowledge several people and organizations that have been critical to the reporting and writing of this book.

First the seed: Several decades ago, the co-founders of the Center for Investigative Reporting, Dan Noyes and David Weir, showed faith in a young journalist and encouraged me to go to Iowa to track down the impacts of genetic uniformity on American corn fields. That experience, encouraged also by one of my first magazine editors, and now friend, Mark Dowie, planted the ideas that lingered and fermented over many intervening years to become this book.

Thank you to those whose financial support for my recent work has enabled me to pursue much of the reporting for this book. Foremost among them are Maggie Kaplan, President and Founder of Invoking the Pause, a small fund with big impact; and Sarah Bell at the 11th Hour Project, which funded some of the initial stages of reporting on the impact of climate change on agriculture that is a backbeat to this book. Much gratitude, also, to the Blue Mountain Center, which provided me a fine place to work on the early chapters alongside the lovely Blue Mountain Lake in the Adirondacks. And to the Mesa Refuge, whose beautiful settings I've been fortunate enough to avail myself over the past several years.

Thank you to my friend David Talbot, professional provocateur and editor of this series who first suggested I do this book for Hot Books, and played a key role in bringing it to publication. Thanks to the fine people at Skyhorse for their support for this project, notably Mark Gompertz, Caroline Russomanno, and Tony Lyons. My agent, Diana Finch, went beyond the call of duty to offer valuable editorial advice and critiques. Thank you James Wheaton, who provided pro bono legal advice. Thanks to Heidi Quant for her helpful comments along the way—and to my friends who absorbed my endless riffs on seeds over several years. And thanks to the many multiple generations of people on every corner of the planet who have mastered the art of conjuring food from the earth.

Finally, much gratitude to my family—to my brother Seth Schapiro, amd Abbey Asher-Schapiro, Avi Asher-Schapiro, Shayna Asher-Schapiro, Lani Asher, and Stephen Kamelgarn for being sources of emotional sustenance and joy. My brother Erik Schapiro passed away far too early while I wrote this book, and his presence, his love of the natural world, his generous spirit, and his humor runs through these pages.

Finally, thank you to my partner, Zoe Fitzgerald Carter, for her love and support, for her many insights, some very sharp edits and inspiration from the music she made in the other room; and for her understanding of what it means to struggle with a sentence, and what it means to sustain a rhythm.

# NOTES

## INTRODUCTION

1. Mark Schapiro, "Seeds of Disaster," *Mother Jones*, December 1982.
2. Jane E. Brody, "Tests Find Stimuli Aid Brain Growth," January 23, 1970, *New York Times*. This article was based on a series of scientific findings published in 1969-1970 in the journal *Physiology and Behavior*, by Shawn Schapiro and Manuel Salas.
3. A classic book on the evolving history of scientific paradigms provides continuing inspiration on the links between the scientific method and journalism: Thomas S. Kuhn, *The Structure of Scientific Revolutions*, Third Edition, (Chicago, IL: University of Chicago Press, 1996).
4. Daniel Chamowitz, *What a Plant Knows: A Field Guide to the Senses*, Updated and Expanded Edition (Scientific American/Farrar, Straus and Geroux, 2017). (Collects in one book much of the current knowledge about the responsiveness of plants to their environment).

## CHAPTER 1

1. A.J. Challinor, et al, "Current Warming Will Reduce Yields Unless Maize Breeding and Seed Systems Adapt Immediately," *Nature Climate Change*, June 20, 2016.

2. Louise Sperling, et al, "Moving Towards More Effective Seed Aid," *Journal of Development Studies*, vol. 44, April 2008.; A primer for setting up a seed bank: "Defining Seed Quality and Principles: Seed Storage in a Smallholder Context, Brief No. 1," by Catholic Relief Services for the U.S. Agency for International Development Office of Foreign Disaster Assistance.

3. A history of the Svalbard Seed Vault written by its founder: Cary Fowler, *Seeds on Ice: Svalbard and the Global Seed Vault* (Prospecta Press, 2016).

4. Xin-Zhong Liang, You Wu, Robert G. Chambers, Daniel L. Schmoldt, et al, "Determining Climate Effects on US Total Agricultural Productivity," *Proceedings of the National Academy of Sciences* (PNAS), March 6, 2017.

5. "Climate Change, Global Food Security and the U.S. Food System," US Department of Agriculture, December 2015, "Executive Summary."

6. David B. Lobell and Christopher B. Field, "California Perennial Crops in a Changing Climate," *Climatic Change*, November 2011.

7. Eliza Roberts and Brooke Barton "Feeding Ourselves Thirsty: How the Food Sector is Managing Global Water Risks," *Ceres*, May 2015.

8. "Action Against Desertification, Background," Food and Agriculture Organization: http://www.fao.org/in-action/action-against-desertification /background/en/.

9. "Climate Change, Global Food Security and the U.S. Food System," US Dept of Agriculture, December 2015.

10. Oliver Milman, "US Federal Department is Censoring Use of Term 'Climate Change', Emails Reveal," *The Guardian*, August 7, 2017.

## CHAPTER 2

1. I first met McLain and his family in 1982 while reporting a story on seed industry consolidation, then in its early stages. The interview of Fred McLain and details from the McLain's farm are drawn from the article I published: "Seeds of Disaster," *Mother Jones*, December 1982.

2. "Genetic Vulnerability of Major Crops," National Research Council-Committee on Genetic Vulnerability of Major Crops, National Academy of Sciences, 1972.

3. "Genetic Vulnerability of Major Crops," National Research Council-Committee on Genetic Vulnerability of Major Crops, National Academy of Sciences, 1972.

4. Kiersten Wise, "Gray Leaf Spot," *Diseases of Corn* series, *Purdue Extension*, Purdue University.

5. For those familiar with Arabic, some of the stories were published in association with the Arab Network for Investigative Journalists (ARIJ): www.arij.net.

6. Francesca De Chatel, "The Role of Drought and Climate Change in the Syrian Uprising: Untangling the Triggers of the Syrian Revolution," *Journal of Middle Eastern Studies*, Volume 50, 2014, issue 4.; Peter H. Gleick, "Water, Drought, Climate Change, and Conflict in Syria," *American Meteorological Society*, July 2014.

7. The Ry Cooder-Ali Farka Toure collaboration is enshrined in their extraordinary album *Talking Timbuktu*, which won the Grammy for Best World Music Album of 1994.

8. John W. Troutman, *Kika Kika: How the Hawaiian Steel Guitar Changed the Sound of Modern Music* (University of North Carolina Press, 2016).

9. Gary Paul Nabhan, *Where Our Food Comes From: Retracing Nikolay Vavilov's Quest to End Famine*, (Washington, DC: Island Press, 2011).

10. Ibid.

11. Maywa Montenegro, "Banking on Wild Relatives to Feed the World," *GASTRONOMICA: The Journal of Critical Food Studies*, March 2017.

12. Colin Khoury, "We All Benefit from Foreign Nations' Food Crop Diversity—But Do Our Politics Reflect this Interdependence?" Union of Concerned Scientists, May 30, 2017.

13. A trove of biographical information on Luther Burbank is contained in: Jane S. Smith, *The Garden of Invention: Luther Burbank and the Business of Breeding Plants* (Penguin Press HC, 2009).

14. Peter Dreyer, *A Gardener Touched With Genius: The Life of Luther Burbank* (Santa Rosa, CA: Luther Burbank Home and Gardens, 2002). Includes the 101-page catalog of "Luther Burbank's Plant Contributions."

15. James Crow, "Plant Breeding Giants: Burbank, the Artist; Vavilov, the Scientist," *Genetics*, August 2001.

16. Igor G. Loskutov, "Vavilov and His Institute: A history of the world collection of plant genetic resources in Russia," *International Plant Genetic Resources Institute*, 1993, 19-20.

17. James F. Crow, "Plant Breeding Giants: Burbank, the Artist; Vavilov, The Scientist," *Genetics*, August 2001.

18. "Genetic Vulnerability of Major Crops," National Academy of Sciences, Division of Biology and Agriculture, National Research Council, 1972.

19. Peter Pringle, *The Murder of Nikolai Vavilov: The Story of Stalin's Persecution of One of the Great Scientists of the Twentieth Century* (New York, NY: Simon & Schuster, 2008).

## CHAPTER 3

1. "Plant Patents: January 1, 1991-December 31, 2015," U.S. Patent and Trademark Office, A Patent Technology Monitoring Team Report, March 2016, https://www.uspto.gov/web/offices/ac/ido/oeip/taf/plant.pdf.

2. Helpful background on the legal history of plant patents: Sheldon Krimsky, "Patents for Life Forms Sui Generis: Some new questions for science, law and society," *Recombinant DNA Technical Bulletin*, April 1981.

3. "Monsanto's Seed Company Subsidiaries: Fact Sheet," Food and Water Watch, June 2014.

4. Philip H. Howard, *Concentration and Power in the Food System: Who Controls What We Eat?* (Bloomsbury Academic, 2015).

5. "Saved Seed and Farmer Lawsuits," Monsanto *News:* https://monsanto.com/company/media/statements/lawsuits-against-farmers/.

6. I am not the first, I discovered, to see this parallel between seeds and high-tech. For more on such parallels, see: Lisa M. Hamilton, "Linux for Lettuce," *Virginia Quarterly Review*, Summer 2014.

7. "Competition and Agriculture: Voices from the Workshops on Agriculture and Anti-Trust Enforcement in our 21st Century Economy and Thoughts on the Way Forward," Issued by the Department of Justice (in coordination with the USDA), May 2012.

8. David Weir and Mark Schapiro, *Circle of Poison: Pesticides and People in a Hungry World*, Institute for Food and Development Policy, 1981.

9. "Seed Relabeling Report," Farmers Business Network, 2017.

10. "What is Agrobiodiversity?" Food and Agriculture Organization; Lori Ann Thrupp, "CULTIVATING DIVERSITY: Agrobiodiversity and Food Security," World Resources Institute, 1998.

11. Jonathan Aguilar, Greta Gamig, et al., "Crop Species Diversity Changes in the United States: 1978-2012," *PLOS One*, August 2015. The scientists based their findings on cultivation records for every agricultural county in the United States.

12. Some of the information about patents in this chapter is drawn from Kloppenberg's book: Jack Kloppenberg, *FIRST THE SEED: The Political Economy of Plant Biotechnology* (University of Wisconsin Press, 2004).

13. "Seeds of the Future: How Investment in Classical Breeding Can Support Sustainable Agriculture," Fact Sheet, Union of Concerned Scientists.

14. "Pesticide Use in U.S. Agriculture: 21 Selected Crops 1960-2008," USDA, Economic Research Service, May 2014.

15. I reported earlier about my experience of returning to the McLain farm for print and public television: Mark Schapiro, "SOWING DISASTER: How Genetically Engineered American Corn Has Altered the Global Landscape," *The Nation*, October 10, 2002.; "Seeds of Conflict," *NOW With Bill Moyers*, aired nationally on PBS stations, October 4, 2002.

16. William Freese and David Schubert, "Safety Testing and Regulation of Genetically Engineered Foods," *Biotechnology and Genetic Engineering Review*, vol. 21, 2004.

17. Kurt Eichenwald, Gina Kolata and Melody Petersen, "Biotechnology Food: From the Lab to a Debacle," *New York Times*, January 25, 2001.

## CHAPTER 4

1. Catherine Green, Seth Wechsler, et al, "Economic Issues in the Coexistence of Organic, Genetically Engineered (GE) and Non-GE Crops," US Department of Agriculture, Economic Research Service Bulletin, #149, February 2016. Additional information: "Organic Farmers Pay the Price for GMO Contamination," Food & Water Watch and Organic Farmers Agency for Relationship Marketing (OFARM), Issue Brief.

2. "GENETICALLY ENGINEERED CROPS: USDA Needs to Enhance Oversight and Better Understand Impacts of Unintended Mixing with Other Crops," Government Accountability Office, Report to U.S. Senator Jon Tester, March 2016.

3. "Organic Farmers Pay the Price for GMO Contamination," Issue Brief, Food & Water Watch and O-FARM.

4. "A Framework for Local Coexistence Discussions," A Report of the Advisory Committee on Biotechnology and 21st Century Agriculture (AC21) to the Secretary of Agriculture, December 8, 2016.

## CHAPTER 5

1. "Adoption of Genetically Engineered Crops in the US," Economic Research Service, US Department of Agriculture, http://www.ers.usda.gov/data -products/adoption-of-genetically-engineered-crops-in-the-us.aspx.

2. International Service for the Acquisition of Agri-Biotech Applications, "ISAAA Brief 51-2015, Executive Summary," http://www.isaaa.org /resources/publications/briefs/51/executivesummary/default.asp.

3. Libby Foley, "Big Food Companies Spend Millions to Defeat GMO Labeling," Environmental Working Group, August 4, 2015.

4. Kari Hamerschlag, Anna Lappé and Stacy Malkan, "SPINNING FOOD: How Food Industry Front Groups and Covert Communications are Shaping the Story of Food," Friends of the Earth.

5. OpenSecrets.org, Center for Responsive Politics, Kansas District 04, Mike Pompeo.

6. Thomas Guillemaud, Eric Lombaert et al, "Conflicts of Interest in GM Bt Crop Efficacy and Durability Studies," *PLOS-One*, December 15, 2016.

7. Sheldon Krimsky and Tim Schwab, "Conflict of Interest Among Committee Members in the National Academies' Genetically Engineered Crop Study," *PLOS-One*, February 28, 2017; "Under the Influence: The National Research Council and GMO's," Issue Brief, Food & Water Watch, May 2016.

8. U.S. Right to Know has unearthed troves of previously hidden documents indicating the breadth of influence of the agri-chemical industries over government agencies and universities. [www.usrtk.org].

9. Sheldon Krimsky, "An Illusory Consensus Behind GMO Health Assessment," *Journal of Science, Technology and Human Values*, 2015.

10. Angelika Hilbeck, et al, "No Scientific Consensus on GMO Safety," *Environmental Sciences Europe*, April 2015 (27:4). The article explained the implications of the letter signed by the 300 scientists from around the world.

11. Jonathan Aguilar, Greta Gamig, et al, "Crop Species Diversity Changes in the United States: 1978-2012," *PLOS One*, August 2015.

12. Mark A. Mikel, "Genetic Composition of Contemporary U.S. Commercial Dent Corn Germplasm," *Crop Science*, vol. 51, no.2, 2011.

13. Bruce E. Tabashnik, "ABC's of Insect Resistance to bt," *PLOS One*, November 19, 2015.; Bruce E. Tabashnik, Thierry Brevault, and Tves Carriere, "Insect Resistance to bt Crops: Lessons from the First Billion Acres," *Nature Biotechnology*, June 2013.; Yunxin Huang, Yun Qin, et al, "Modelling the Evolution of Insect Resistance to One and Two-Toxin bt Crops in Spatially Heterogenous Environments," *Ecological Modelling*, March 2017.; "Scientists Discover How Bollworm Becomes Resistant to bt Crops," *Entomology Today*, News from the Entomological Society of America, March 13, 2015.

14. International Survey of Herbicide Resistant Weeds: http://www.weedscience.org.

15. "The Rise of Superweeds—and What to Do About It," Union of Concerned Scientists Policy Brief.

16. Margaret Worthington and Chris Reberg-Horton, "Breeding Cereal Crops for Enhanced Weed Suppression: Optimizing Allelopathy and Competitive Ability," *Journal of Chemical Ecology*, January 2013.

17. "Genetically Engineered Crops: Experiences and Prospects," Board on Agriculture and Natural Resources, National Academy of Sciences, May 2016.

18. Danny Hakim, "Doubts About the Promised Bounty of Genetically Modified Crops," *New York Times*, October 29, 2016.

19. Jack A. Heinemann, et al, "Sustainability and Innovation in Staple Crop Production in the US Midwest," *International Journal of Agricultural Sustainability*, issue 1, vol. 12, 2014.

20. "Restrictions on Genetically Modified Organisms: New Zealand," (U.S.) Library of Congress, The Law Library, https://www.loc.gov/law/help /restrictions-on-GMOs/new-zealand.php#_ftn1.

21. "Trends in Glyphosate Herbicide Use in the United States and Globally," by Charles M Benbrook, *Environmental Sciences Europe*, March 2016.

22. Monsanto Press Release announcing 2016 Annual Report: http://news .monsanto.com/press-release/financial/continued-soybean-technology -expansion-and-cost-discipline-expected-drive-re.

23. Maxx Chatsko, "How Much Money Does Monsanto Make from Roundup?" *The Motley Fool* (investor advice newsletter, stats based on SEC filings), May 26, 2016.

24. "United States of America Before the Securities and Exchange Commission, in the Matter of Monsanto Company, Sara M. Brunnquell, CPA, Anthony P Hartke, CPA and Jonathan W. Nienas," Accounting and Auditing Enforcement, Release No. 3741, Administrative Proceeding File No. 3-17107," February 9, 2016.

25. Philip H. Howard, *Concentration and Power in the Food System: Who Controls What We Eat?* (Bloomsbury Academic, 2016).

26. "IARC Monographs Volume 112: Evaluation of Five Organophosphate Pesticides and Herbicides," International Agency for Research on Cancer, March 20, 2015.; Daniel Cressey, "Widely Used Herbicide Linked to Cancer: The World Health Organization's research arm declares glyphosate a probable carcinogen. What's the Evidence?" *Nature*, reprinted by *Scientific American*, March 25, 2015.

27. Office of Environmental Health Hazard Assessment, California Environmental Protection Agency, "The Proposition 65 List," http://oehha.ca.gov /proposition-65/proposition-65-list.

28. "Monsanto Company's Memorandum of Points and Authorities in Opposition to OEHHA's Motion for Judgment on the Pleadings and the Sierra Club Intervenors' Demurrer," Filed with Superior Court of California, County of Fresno, December 9, 2016 as part of documents submitted by Monsanto in the case of Monsanto Company and California Citrus Mutual vs. Office of Environmental Health Hazard Assessment, Et Al and Sierra Club, Center for Food Safety, Et Al.

29. Kate Kellund, "Cancer Agency Left in the Dark Over Glyphosate Evidence," *Reuters Investigates*, June 14, 2017, https://www.reuters.com/investigates /special-report/glyphosate-cancer-data/.

30. Carey Gilliam and Stacy Malkan, "Reuters Kate Kelland Story Promotes False Narrative," U.S. Right To Know, July 24, 2017: https://usrtk.org /pesticides/reuters-kate-kelland-iarc-story-promotes-false-narrative/.

31. "The Monsanto Papers, Part 1: Operation Intoxication," and "The Monsanto Papers, Part 2: Reaping a Bitter Harvest," translation into English from the French by the Health and Environment Alliance, November 20 and 21, 2017 [originally published in French in *Le Monde* on June 1 and 2, 2017].

32. Carey Gilliam, *Whitewash: The Story of a Weedkiller, Cancer and the Corruption of Science* (Washington, DC: Island Press, 2017).

33. Joel Rosenblatt, Lydia Mulvaney and Peter Waldman, "EPA Official Accused of Helping Monsanto 'Kill' Cancer Study," *Bloomberg Markets*, March 14, 2017. Many of these documents were unearthed by the NGO U.S. Right to Know, and can be seen at: "Monsanto Glyphosate Cancer Case Key Documents & Analysis," by Gary Ruskin, posted March 14, 2017.

34. Peter Waldman, Tiffany Stecker and Joel Rosenblatt, "Monsanto Was Its Own Ghostwriter for Some of its Safety Reviews," *Bloomberg Business Week*, August 9, 2017.

## CHAPTER 6

1. Hazem Badr, "Syria's ICARDA falls to rebels, but research goes on," *SciDevNet*, May 22, 2014.

2. "State Annual and Seasonal Time Series," National Climatic Data Center, National Oceanic and Atmospheric Administration, https://www.ncdc .noaa.gov/temp-and-precip/state-temps/.

3. I wrote about this dispute and seed libraries in: "Seed Librarians at the Front Lines," *Pacific Standard*, March-April 2017.

4. Frederik van Oudenhoven and Jamila Haider, *With Our Own Hands: A Celebration of Food and Life in the Pamir Mountains of Afghanistan and Tajikistan*, (Netherlands: LM Publishers and University of Washington Press, 2015).

5. H. Muminjanov, M. Otambekova and A. Morgounov, "The History of Wheat Breeding in Tajikistan," *World Wheat Book*, Volume 3. A History of wheat breeding, 2016.

## CHAPTER 7

1. "Assessment of Climate Change in the Southwest United States," *A Report Prepared for the National Climate Assessment, National Climate Assessment Regional Technical Input Report Series* (Washington, DC: Island Press, 2013).
2. Nabhan, *Where Our Food Comes From.*
3. Steven Brescia, ed. *Fertile Ground: Scaling Agroecology from the Ground Up* (Food First Books, 2017).
4. "Sorting Out the Facts Behind Stacks," Monsanto Corporation website: https://monsanto.com/innovations/biotech-gmos/articles/agriculture -stacks/.
5. Thin Lei Win, "Eco-farming Can Solve Hunger and Climate Crisis, Experts Say," Thompson Reuters Foundation, April 4, 2018; "FAO's Work on Agro-Ecology: A Pathway to Achieving the SDG's," Food and Agriculture Organization.
6. "Sequestering Carbon in Soil: Addressing the Climate Threat," Summary Report, Breakthrough Strategies and Solutions, August 2017; Moises Velasquez-Manoff, "Can Dirt Save the Earth?" *New York Times Magazine*, April 22, 2018; Jacques Leslie, "Soil Power! The Dirty Way to a Green Planet," *New York Times Sunday Review*, December 2, 2017.
7. Anna Lappe's book, *Diet for a Hot Planet: The Climate Crisis at the End of Your Fork and What You Can Do About It* (Bloomsbury, 2010), proposes ecological and climate-sensitive practices for growing our food.
8. "Sequestering Carbon in Soil: Addressing the Climate Threat," Breakthrough Strategies and Solutions, August 2017; Jacques Leslie, "Soil Power! The Dirty Way to a Green Planet," *New York Times Sunday Review*, December 2, 2017.
9. Claire Kremen, Alastair Iles and Christopher Bacon, "Diversified Farming Systems: An Agroecological, Systems-based Alternative to Modern Industrial Agriculture," *Journal of Ecology and Society*, vol. 17, 2012.

10. Rita Seidel, Jeff Moyer, et al, "Studies on Long-Term Performance of Organic and Conventional Cropping Systems in Pennsylvania," *Journal of Organic Agriculture*, December 2015.

11. Andrew R. Kniss, Steven D. Savage, and Randa Jabhour, "Commercial Crop Yields Reveal Strengths and Weaknesses for Organic Agriculture in the United States," *PLOS-One*, August 23, 2016.; Lauren C. Ponisio and Paul R. Ehrlich, "Diversification, Yield and New Agricultural Revolution: Problems and Prospects," *Sustainability*, November 2016.

12. Jonathan G. Lundgren and Scott W. Fausti, "Trading Biodiversity for Pest Problems," *ScienceAdvances*, AAAS, July 31, 2015.

13. Miguel Altieri, Clara Nicholls, et al, "Agroecology and the Design of Climate Change-Resilient Farming Systems," *Journal of Agronomy and Sustainable Development*, May 1, 2015.

14. "Study Finds Climate Change May Dramatically Reduce Wheat Production," February 18, 2015, http://www.k-state.edu/media/newsreleases/feb15/climatewheat21815.html.

15. Jonathan Aguilar, Greta Gramig, and John Hendrickson, "Crop Species Diversity Changes in the United States: 1978-2012," *PLOS-One*, August 26, 2015.

16. USDA Risk Management Agency, Data Indemnity Maps, https://www.rma.usda.gov/data/indemnity/2016/.

17. "High Risk Series: An Update," Report to Congressional Committees, Government Accountability Office, 2013 and 2015.

18. Anne Weir Schechinger and Craig Cox, "Is Federal Crop Insurance Policy Leading to Another Dust Bowl?" Environmental Working Group, March 2017.

19. "Climate Change and Agriculture in the United States: Effects and Adaptation," US Department of Agriculture, February 2013.

20. I originally wrote about Ceccarelli for the Food and Environmental Reporting Network (FERN): "Syrian Seeds Shake Up Europe's Plant Patent Regime," *FERN*, September 13, 2017.

## CHAPTER 8

1. www.bio.org.

2. "The Global Farmland Grab in 2016: How Big, How Bad?" GRAIN.org, June 2016. An ongoing record of land acquisitions by foreign companies and governments in Africa, Asia and Latin America is maintained by the independent research organization The Land Matrix: http://www .landmatrix.org/en/. And a study of the phenomenon: Maria Cristina Rulli, et al, "Global Land and Water Grabbing," Proceedings of the National Academy of Sciences, December 2012.

3. Sonia Narang, "Why are Kids Going Hungry in one of California's Most Productive Farming Regions?" PRI's *The World*, January 28, 2016; Also see 2015 Annual Report of Feeding American (which runs the largest network of food banks in the United States): http://www.feedingamerica.org /hunger-in-america/our-research/map-the-meal-gap/2013/CA _AllCounties_CDs_MMG_2013.pdf.

4. Jonathan Foley, "It's Time to Rethink America's Corn System," *Scientific American*, March 5, 2013.

5. "Corn and Other Feed Grains," USDA Economic Research Service, https:// www.ers.usda.gov/topics/crops/corn/background.aspx.

6. "Food Security in the U.S., Key Statistics," USDA Economic Research Service, https://www.ers.usda.gov/topics/food-nutrition-assistance/food -security-in-the-us/key-statistics-graphics.aspx#foodsecure.

7. "The State of Food Insecurity in the World-2015," Food and Agriculture Organization of the United Nations, International Fund for Agricultural Development and the World Food Program, 2015.

8. "Report of the Special Rapporteur on the Right to Food," United Nations Human Rights Council, 34th session, January 24, 2017.

9. Ibid.

10. P.C.D. Milly, Julio Betancourt, et al, "Stationarity is Dead: Whither Water Management," *Science*, February 1, 2008.; P.C.D. Milly, Julio Betancourt, et al, "On Critiques of 'Stationarity is Dead: Whither Water Management'," *Water Resources Research*, American Geophysical Union, September 1, 2015.

11. I wrote a book exploring the economic and political implications of this volatility: *The End of Stationarity: Searching for the New Normal in the Age of Carbon Shock* (Chelsea Green, 2016). That book also includes the quote earlier in the chapter from Christopher Milly.

12. For this George Bernard Shaw quote, thanks to the book: Thor Hanson, *The Triumph of Seeds: How Grains, Nuts, Kernels, Pulses & Pips Conquered the Plant Kingdom and Shaped Human History* (Basic Books, 2016).

# INDEX

2,4-D, 82

Abu Ghraib, Iraq, 93–94, 100
AC-21, 67–68
Academic Reviews, 74
Advisory Committee on Biotechnology
    and 21st Century Agriculture,
    67–68
Africa, 123
Agent Orange, 43, 82
agricultural chemicals, 10
agriculture, beginning of, 1–2
alfalfa, 77
al Hamka, Hussein, 100–101
Altieri, Miguel, 133–134
American Seed Trade Association, 34
Amri, Ahmed, 96, 99, 149
Arab Network for Investigative
    Journalism (ARIJ), 26
*Arche Noah*, 140
Arizona, 16–17, 111, 114, 116–122
Arkansas, 82
artificial selection, 2, 14, 77
Asgrow, 43, 45
Australia, 17

*Bacillus thuringiensis*, 79
bacteria, 42
banana, 32

*Banatka*, 100
Baron, Charles, 50
BASF, 45
Battir, 97
Bayer, 45, 50
Bayer-Monsanto, 10, 148
beans, 32
Biotechnology Industry Organization,
    72, 83, 143
Biotechnology Risk Assessment
    Program, 73
Brascoupe, Clayton, 119
Brummer, Charles, 54–55
Bt, 79, 81
Burbank, Luther, 33–38, 77
Burbank Experimental Farm and
    Orchard, 34
Bureau of Plant Industry, 104
Burkina Faso, 126
Burpee, W. Atlee, 39
Bush, George H.W., 57

California, 16–17, 72, 75, 88, 134, 144
canola, 77, 84
Carson, Rachel, 3, 129
Carver, George Washington, 39
Ceccarelli, Salvatore, 55, 95, 97–98,
    136–137, 139–141, 149
Center for Food Safety, 79

# INDEX